This Man's Pill

This Man's Pill

Reflections on the 50th Birthday of the Pill

Carl Djerassi

OXFORD
UNIVERSITY PRESS

Great Clarendon Street, Oxford OX2 6DP

Oxford University Press is a department of the University of Oxford.
It furthers the University's objective of excellence in research, scholarship,
and education by publishing worldwide in

Oxford New York

Athens Auckland Bangkok Bogotá Buenos Aires Calcutta
Cape Town Chennai Dar es Salaam Delhi Florence Hong Kong Istanbul
Karachi Kuala Lumpur Madrid Melbourne Mexico City Mumbai
Nairobi Paris São Paulo Singapore Taipei Tokyo Toronto Warsaw

with associated companies in Berlin Ibadan

Oxford is a registered trade mark of Oxford University Press
in the UK and in certain other countries

Published in the United States
by Oxford University Press Inc., New York

First published 2001

British Library Cataloguing in Publication Data
Data available

Library of Congress Cataloguing in Publication Data

1 3 5 7 9 10 8 6 4 2

ISBN 0-19-850872 7 (Hbk)

Typeset in Minion by Florence Production, Stoodleigh, Devon
Printed in Great Britain on acid-free paper by
T.J. International Ltd, Padstow

Demanding the perfect is the
enemy of achieving the possible

Author's Biographical Sketch

Carl Djerassi, born in Vienna but educated in the US, is a writer and professor of chemistry at Stanford University. Author of over 1200 scientific publications and seven monographs, he is one of the few American scientists to have been awarded both the National Medal of Science (in 1973, for the first synthesis of a steroid oral contraceptive—'the Pill') and the National Medal of Technology (in 1991, for promoting new approaches to insect control). A member of the US National Academy of Sciences and the American Academy of Arts and Sciences as well as many foreign academies, Djerassi has received 18 honorary doctorates together with numerous other honors, such as the first Wolf Prize in Chemistry, the first Award for Industrial Application of Science from the National Academy of Sciences, and the American Chemical Society's highest award, the Priestley Medal. He is also an honorary member of the Royal Society of Chemistry.

For the past decade, he has turned to fiction writing, mostly in the genre of 'science-in-fiction,' whereby he illustrates, in the guise of realistic fiction, the human side of scientists and the personal conflicts faced by scientists in their quest for scientific knowledge, personal recognition, and financial rewards. In addition to novels (*Cantor's Dilemma*; *The Bourbaki Gambit*; *Marx, deceased*; *Menachem's Seed*; *NO*), short stories (*The Futurist and Other Stories*), and autobiography (*The Pill, Pygmy Chimps, and Degas' Horse*)—most of them written in London—he has recently embarked on a trilogy of plays with an emphasis on contemporary cutting-edge research in the biomedical sciences, which he describes in his web site as 'science in theater.' *An Immaculate Misconception*, first performed in abbreviated form at the 1998 Edinburgh Fringe Festival and subsequently (1999) as a full, two-act play in London (New End Theater), San Francisco (Eureka Theater) and Vienna (under the title *Unbefleckt* at the Jugendstiltheater), focuses on the ethical issues inherent in recent spectacular advances in the treatment of male infertility through single sperm injection (the ICSI technique). A radio adaption was broadcast over the BBC World Service as 'Play of the Week' in 2000 and is scheduled for 2001 by the West German Radio as are French and Bulgarian theater premieres. His second play, *Oxygen*, written with Roald Hoffmann, premiered in April 2001 at the San Diego Repertory w. The BBC World Service has selected it for broadcast in 2001 as 'Play of the Week'.

He is also the founder of the Djerassi Resident Artists Program near Woodside, California, which provides residencies and studio space for artists in the visual arts, literature, choreography and performing arts, and music. Over 1000 artists have passed through that program since its inception in 1982.

(There is a web site about Carl Djerassi's writing at **http://www.djerassi.com**)

By the same author

Fiction

The Futurist and Other Stories

Cantor's Dilemma

The Bourbaki Gambit

Marx, Deceased

Menachem's Seed

NO

Poetry

The Clock Runs Backward

Drama

An Immaculate Misconception

Oxygen (with Roald Hoffmann)

Non-fiction

The Politics of Contraception

Steroids Made it Possible

The Pill, Pygmy Chimps, and Degas' Horse

From the Lab into the World: A Pill for People, Pets, and Bugs

Scientific monographs

Optical Rotatory Dispersion: Applications to Organic Chemistry

Steroid Reactions: An Outline for Organic Chemists (editor)

Interpretation of Mass Spectra of Organic Compounds (with H. Budzikiewicz and D.H. Williams)

Structure Elucidation of Natural Products by Mass Spectrometry (2 volumes with H. Budzikiewicz and D.H. Williams)

Mass Spectrometry of Organic Compounds (with H. Budzikiewicz and D.H. Williams)

For Diane Middlebrook,
"La Ultima"

Contents

1 An exaltation of thirty: Murasaki and company *1*

2 Genealogy and birth of the Pill *11*

3 Bitter Pills *63*

4 The view from Tokyo *98*

5 Sex and immortality *114*

6 From the Pill to the PC *137*

7 Science-in-fiction is not science fiction. Is it autobiography? *151*

8 Behind the scrim of fiction *168*

9 The softer chemist *188*

10 The Pill and Paul Klee *214*

11 Science on stage *244*

12 What if? *280*

Chapter 1

An exaltation of thirty: Murasaki and company

In the cover story of its 12 September 1999 issue, the London *Sunday Times Magazine* featured The Top Thirty persons of the present millennium. Generated by 15 British and American scholars, this list was, to put it mildly, idiosyncratic, if not outright bizarre; in other words, one more manifestation of end-of-millennium hoopla. Arranged in chronological order, the list started with the name of the only woman among this androgenic crowd. Although an engaging choice, her appearance owed, I suspect, more to political correctness and one-upmanship than logic. I cannot imagine a popular vote—even in Japan—that would have propelled Murasaki Shikibu into that 'Gang of Thirty'. But leading with her blunted not only the unavoidable accusation of male bias, but also that of Eurocentrism: Asia was represented only by Murasaki Shikibu and the Ottoman Sultan Mohammed II. Genghis Khan, Mahatma Gandhi and Mao Tse-tung were nowhere in sight. Why did Rupert Murdoch's experts select Napoleon and Lenin, but none of those Asian leaders? The short shrift given the Americas, from Baffin Bay all the way down to Patagonia, showed only that the panel's ethnocentrism was even-handed. No Cortez, Bolivar, Washington, Lincoln, or Roosevelt; indeed only one all-American entry: Thomas Alva Edison. Africa, of course, remained to these intellectual explorers a dark continent.

But the exclusions of geography and gender were at least comprehensible; the more striking feature of the list was its sheer capriciousness. The most glaring omissions were the musicians —no Bach, no Mozart, no Verdi, not even the Beatles—and no artists other than Leonardo da Vinci, whose inclusion seemed to be based primarily on his credentials as a scientist and engineer. The choices in literature were somewhat better: Aside from Murasaki, there was the toweringly obvious choice of Shakespeare, and the almost equally inevitable ones of Dante and Chaucer. Jean Jacques Rousseau, the remaining literary entry, seemed unexceptionable enough, but only until one thought to ask why him, and not, say, Goethe or Tolstoy.

The list was heavily skewed in favor of science and technology. Among the 15 scientists and inventors were Bacon, Newton, Copernicus, Galileo, Darwin, Pasteur, and Einstein. All of them seem rational choices, although some alternative ones—say Planck, Maxwell, Watson and Crick (probably eliminated for cannibalizing a precious pair of vacancies)—would have been just as plausible. The British bias was glaringly obvious, but why not? After all, this was *The Times* of London, not the *New York Times.* And of the 15 'expert judges,' 11 came from the UK and only four from the States. So there was little surprise in the final vote for *primus inter pares*—the *numero uno* of the millennium:

Isaac Newton. The only irony was that the cover announcing the choice did not feature a visually hagiographic portrait of the physicist (of which there are plenty), but rather a photograph of Eduardo Paolozzi's giant sculpture of Newton at the entrance to the British Library, a sculpture based on William Blake's famous 1795 watercolor. Did *The Sunday Times*'s photo editor know that Blake's views of Newton and his rationalism were hardly complimentary? 'He who sees the infinite in all things sees God. He who sees the Ratio only sees himself only.'

The Sunday Times's vote is meaningless, of course. There cannot be *one* most important person of the millennium: no single criterion or even set of criteria could ever gain general acceptance. There is no longer (if there ever was) a single community of learned men or women that such a figure or a list could represent. Which brings me to the point—most certainly not original with me—that Newton, Galileo, Einstein, and all the other scientific savants do not appear on that list as *persons*—in the sense that Murasaki and Shakespeare do—but rather as representatives, surrogates for discoveries or inventions.

Newton's greatest revelations, such as the laws of gravity and of motion, might have taken a few more years to be discovered if he had never been born—but discovered they inevitably would have been—as Leibnitz's concurrent invention of the calculus—but one example of the simultaneous discoveries that litter the history of science—so amply demonstrates. Doubtlessly, Copernicus, Galileo, Darwin, and Einstein were the first with their breakthrough insights in their respective fields, but again, somebody else would have come up with the same generalizations within a time frame that can only be termed insignificant on the scale of human history. In the final analysis, in science, unlike art, the individual hardly matters.

Well, hardly ever. Unless the individual happens to be oneself. *The Sunday Times*'s list ends with one living relic. On the face of it, the appearance of the name Carl Djerassi is patently ludicrous by any criterion but one: as a surrogate for the Pill. Few would question that during the past four decades of this millennium the introduction of steroid oral contraceptives has had a huge effect. Much, though certainly not all of it, was beneficial. Other medical inventions—say X-rays or antibiotics—have touched greater numbers of people, although even in that regard, the Pill is no lightweight: in the US, 80 per cent of all women born since 1945 have used it! But in terms of socio-cultural impact, ranging from religion to women's rights, the Pill must surely rank close to the top. By separating the coital act from contraception, the Pill started one of the most monumental movements in recent times, the gradual divorce of sex from reproduction. The subsequent advent of in-vitro fertilization techniques has made the complete separation of sex and reproduction—in other words, the creation of new life without sexual intercourse—a reality. The ethical and social implications of this uncoupling are huge, and are only now starting to be debated. While the introduction of the Pill started that reproductive revolution, the parents of the Pill could hardly have anticipated all the consequences arising during the middle age of their 50-year-old offspring.

'Fifty years?' you may ask. When, precisely, was the Pill's birthday? And where was it born? Or, even more to the point, where was it conceived? Most of the time, parents of real babies have difficulties answering that last question, so how to answer it in this instance, where even the identity of the parents is often questioned? Yet the answer is straightforward. The idea of a Pill permitting sex without fertilization arose in the 1920s in Austria. I was not even born at the time of the Pill's conception, yet I will argue that as an organic chemist, I played a maternal role in the birth of the Pill, in Mexico City on 15 October 1951.

The present account is less one of maternal chemistry, however, than it is an evaluation of the effect that the offspring—the Pill—had on the world around us, and especially on me. The story is as yet unfinished, because contrary to everyone's expectation—and especially that of the Pill's parents—contraception has hardly changed during the past 50 years, and seems unlikely to change in any fundamental technical way for at least several more decades. In fact, we may be on the brink of seeing 'contraception' gradually fade away as an issue as we confront a very different future for human reproduction.

But on a purely personal level, the Pill has also had a monumental effect on me. It has changed me from a 'hard' scientist—an organic chemist, driven by scientific curiosity and the scientist's accompanying baggage of ambition, competition, and the desire for peer recognition of which I have written extensively in my novels—to a 'softer' one. I became progressively more occupied with questions actually tougher and more ambiguous than the challenge of corralling carbon atoms into hitherto unknown and often useful forms: the social consequences arising from scientific and technological developments. I credit my final jump, to fiction writing and then drama, to my search for new ways of communicating scientific thoughts and problems to a wider public. It helped that I had the great luck to be associated with that crucial invention, while still in my 20s, so that half a century later I can still reflect on what that discovery has done to me. Hence the title, *This Man's Pill*, is not a statement of proprietorship—even less of braggadocio or naïve macho conceit—but rather the distillate of a self-examination that is far from over.

The question might well be asked: why should I find it necessary to share these conclusions with others? Why not be content with having learned something about myself? Or, if the urge to talk about these issues is so overpowering, why not just see a shrink? After all, a decade ago, I published a collection of memoirs (*The Pill, Pygmy Chimps, and Degas' Horse*) which was as autobiographical a record as my residual sense of privacy allowed.

My answer is simple: I am now ten years older, and to that extent ten years wiser. At heart, I am a pedagogue, who believes that there is plenty to be learned from my life—examples both positive and negative. For instance, in this increasingly geriatric world of ours, it is useful to demonstrate that one can start new careers at age 60 and again at 75. My laboratory is now closed. I am freed to reflect on past events through the filter and haze of over half a century, as well as from a distance that permits a look at the future with the benefit of a wider vision. For me, this book has turned into a form of public penance for sins of past omissions. For having been too busy as a scientist to spend much time communicating with a broader public; too occupied with analyzing the world at large to apply the finely honed analytical skills of a scientist to self-reflection. What better time to do so than on the fiftieth anniversary of the birth of the Pill?

Every scientific paper starts with an abstract. I am too much a creature of habit to break such a fine tradition here. But tradition may bend, and for this personal account I offer as an abstract an autobiography, written in free verse on the sixtieth birthday of the other grandfather, the late Robert Maxwell, of my only grandchild, Alexander Maxwell Djerassi. This abstract-poem has the virtue of being concise as well as brutally honest:

The Clock Runs Backward

At his sixtieth birthday party,
Surrounded by wife, children, and friends,
The man who has everything
Opens his gifts.

Among paperweights, cigars,
Books, silver cases,
Cut glass vases,
Appears a clock
Made by KOOL Designs
In a limited edition.
A clock running backward.
A clock called LOOK.

Amusing.
Just the gift
For the man who has everything.

How Faustian, thought the friend,
Soon to turn sixty himself.
What if it really measured time?

As the hands reached fifty,
He stopped them.
Books, hundreds of papers, dozens of honors.
Not bad, he thought: I like this clock.

But fifty was also the time
His marriage had turned sour.
He let the clock run on.

Forty-eight years, forty-five years,
Then forty-one.
Ah yes, the years of collecting:
Paintings, sculptures, and women.
Especially women.

But wasn't that the time
His loneliness had first begun?
Or was it earlier?
Why else would one collect,
Except to fill a void?

Don't hold the hands!
The thirties were best:
Burst of work. Success. Recognition.
Professor in first-rank university.
Birth of his son—now his only survivor.

What about twenty-eight?
Ah yes—he nearly forgot.
The year of THE PILL.
The pill that changed the world.
No—too pretentious, too self-important.
But he did change the life of millions,
Millions of women taking his pill, he thought.

The clock still regresses.
Twenty-seven years:
First-time father, of a daughter,

In time, his only confessor.
Now dead. Killed herself.
The beginning of his second marriage.
The first undone.

Early stigmata of success to come:
The doctorate not yet twenty-two;
The Bachelor of Arts not yet nineteen.
And the fallacy of presumed maturity:
First-time groom not yet twenty.

Backward: Europe. War.
Hitler. Vienna.
Childhood.
Stop. Stop. STOP!

The pater familias,
Surrounded by wife, children, friends,
The man who has everything
Is still opening presents.
More paperweights, more silver,
More books, ten pounds of Stilton cheese,
And one more clock.

Thank God it's moving forward,
Thought the friend,
The lonely one,
Who'll soon turn sixty himself.

And smiled at the woman at his side,
The one he had met yesterday. Who yesterday had said,

'Yes, I'll come with you to Oslo.'

And come she did.
But not for long.

Chapter 2

Genealogy and birth of the Pill

Wolfson: *What do you understand now that you did not understand when you were 19 or 20?*

Djerassi: *You mean in science, or altogether?*

Wolfson: *Your choice.*

Djerassi: *I don't live in a vacuum. If I had lived those 50 odd years on a desert island, then I think the answer would have been an interior one, but the answer that I have to give you must be based on a reflection of the impact of the world in which I live. The overwhelming fact is that at my birth there were 1.9 billion people in this world. Now there are 5.8 billion and at my 100th birthday, there are likely to be 8.5 billion. That has never before happened in human history—that during a person's lifetime, the world population more than quadrupled. That can never happen again.*

Wolfson: *What have you personally, internally, learned during that period?*

Djerassi: *Internally, I'd have to ask myself the usual question: What would I do differently if I could lead my life over again? My answer would be that I would not lead the same life that I had done before. I would be less of a workaholic. I would be less frenetic. I would simply accept that one couldn't do everything that one wants to do within a lifetime, even though I still have this obsession with not having enough time to do things.* [Pause] *I think it would be interesting to be a woman, now. Being a modern woman could be very interesting because things have changed.*

The above is from a long interview I had in early 1997 with Jill Wolfson, a reporter for the *San Jose Mercury News,* for a series of articles on technical contributions by some Silicon Valley veterans among whom they placed me. Why do I use such an excerpt to start a chapter that contains my views of the origin of the Pill? Because it refers to two overwhelming facts of the past half century—the global population explosion and the rise of women's rights—without which oral contraceptives would just have been another interesting medical advance and not an invention with enormous societal consequences.

I

As I write these lines, we are within months of the fiftieth birthday of the Pill. During the past four years, I have been interviewed, filmed, and encouraged to pontificate on the occasion of the thirty-fifth as well as fortieth birthday of the Pill. How can one celebrate 15 years of birthdays in four years? One of the ironies of the Pill's career is that its own conception has been so hard to pin down. It all depends (as any obstetrician will tell you) on who's counting. In 1997, I spoke at a medical congress in Vienna commemorating the thirty-fifth anniversary of the Pill in Austria—not an inappropriate geographic choice as I will demonstrate shortly—while in May 2000, I was bombarded by requests from many American print, radio, and TV reporters to comment on the 'fortieth birthday' of the Pill. That inquiry had me puzzled until I realized that they were dating the Pill's debut to the formal approval by the Food and Drug Administration (FDA). While such dates may be occasions for celebrations, 'birthdays' they are not. The Viennese event was equivalent to celebrating a baby's arrival in a town far away from its original birthplace and the fortieth 'birthday,' hyped by the American media, could be equated to the date on which the baptismal certificate was issued in Washington. As far as I am concerned (and I *was* concerned), the real birth date of the Pill was 15 October 1951, the day our laboratory completed the first synthesis of a steroid eventually destined to be used for oral contraception. A few days later, the first few precious milligrams of 'norethindrone'—the nickname of the formally named 17α-ethynyl-19-nortestosterone—was already in the mail from the Syntex research laboratories in Mexico City to Dr Elva G. Shipley at Endocrine Laboratories Inc, a commercial establishment in Madison, Wisconsin, with the request that the substance be tested for oral progestational activity.

I mention Dr Shipley here primarily because her early participation in the Pill's history impinges on a claim often heard—that the scientists involved in the development of oral contraceptives were uniformly males. This belief has grated on women for decades. As Margaret Mead put it in 1971: '[The Pill] is entirely the invention of men. And why did they do it? . . . Because they are extraordinarily unwilling to experiment with their own bodies . . . and they're extremely willing to experiment with women's bodies . . . it would be much safer to monkey with men than monkey with women.' While Mead's irritation may well be understandable, it nevertheless represents a gross oversimplification which ignores that nature had provided scientists with a crucial hint on which to build—women do not get pregnant during pregnancy because of the continuous secretion of progesterone—whereas no such clue exists in male reproductive biology. Dr Shipley's contribution is significant for an additional reason that may explain some of Mead's indignation: 50 years ago women were still largely excluded from many areas of scientific research. In a field that was undeniably a male province, Dr Shipley had to do her work in a commercial laboratory she had founded next door to the University of Wisconsin, where her husband was Professor of Zoology, at a time when nepotism rules were still inviolate.

A pervasive sense of the ironies of this historic male bias has caused various writers and journalists to search far and wide for female heroes in the record of the Pill. Margaret Sanger is the favorite, probably for reasons stated in the final paragraph of David Kennedy's definitive 1970 biography:

> Yet the praise Margaret Sanger received often seemed out of proportion to her achievement. Part of the hyperbole, undoubtedly, derived from her personal magnetism, which rarely failed to bring those who met her into her orbit. But a larger part reflected Mrs Sanger's symbolic satisfaction of a

> pervasive psychological need. American society in this century has not realized its frequently stated ideal of equal status for women. Perhaps, therefore, the apotheosization of a feminist heroine like Margaret Sanger reflects society's recognition of the continuing victimization of women, and the desire, in some way, to find a redemptress. For that role Margaret Sanger, at her best and at her worst, was well suited.

Still, Sanger's historic, though certainly not scientific, role in fostering the birth control movement in the USA would validate her choice as one of the grandmothers. A more romanticized candidate is Katherine McCormick, a wealthy philanthropist, who was persuaded by Sanger in the early 1950s to subsidize some of the biological work at the Worcester Foundation for Experimental Biology that, under the leadership of Gregory Pincus, contributed heavily to the development of the Pill. However commendable such philanthropy is, anointing Katherine McCormick as one of 'the indisputable mothers of the Pill' (as was done by one journalistic author, Bernard Asbell, in *A Biography of the Drug that Changed the World* and then repeated by many others), is as far-fetched as calling John D. Rockefeller one of 'the fathers of the Pill.' (The Rockefeller Foundation and its offspring, the Population Council, supported much more research in reproduction and contraception than Mrs McCormick ever did and did so over the course of many decades.) Financial support, valuable as it may be, can never be equated with creativity; otherwise, the Medicis would be considered the greatest artists of the Renaissance. Instead, let me add the name of Elva G. Shipley as literally the first biologist—male or female—who established the high progestational activity of orally-administered norethindrone. If her results had been negative, we would have dropped the project and would never have sent the material to other biologists, notably Roy Hertz and then Gregory Pincus, who, as I will demonstrate below, can rightfully be called a 'father of the Pill.'

Since the Pill is so intimately connected with human reproduction, albeit in terms of preventing it, let me pursue the Pill's genealogy through the metaphor of reproduction. Call the Pill the baby and follow its birth through (1) the first (unsuccessful) attempts at conception, (2) the ovulation of a fertile egg, (3) the ejaculation of various sperm, (4) the successful fertilization, (5) the implantation of the embryo, (6) the fetal development, and finally (7) the birth of the baby. Geographically, the first step occurred in Austria, the second in Mexico, the next three in the continental United States, and the final ones in Puerto Rico—not untypical for a baby in the present highly mobile society.

II

The least known character in the Pill's story is not a woman after all. It is Ludwig Haberlandt, professor of physiology at the University of Innsbruck. As early as 1919, he carried out a crucial experiment, in which he implanted the ovaries of a pregnant rabbit into another rabbit, which, in spite of frequent coitus, remained infertile for several months—a result that Haberlandt called 'hormonal temporary sterilization.' (Partisans of Mrs McCormick might take note that this and subsequent work of Haberlandt's was supported financially by the Rockefeller Foundation.) The problem with this method, of course (other than its reliance on surgery), as well as with subsequent attempts to avoid surgery by the use of 'glandular extracts,' was that these extracts were not the pure hormone responsible for its contraceptive effect. A mixture of hormones and other proteins, they constituted a potential problem of toxicity for the recipient. Attempts to 'purify' these extracts presented the next hurdle to overcome on the way to a practical oral contraceptive.

In numerous subsequent experiments and publications over the course of ten years, Haberlandt—invariably using the first person singular, so strikingly different from today's insistence by scientists on the royal 'we'—emphasized the obvious applicability of his animal experiments to human contraception. He fully recognized that the responsible factor was a constituent of the corpus luteum or 'yellow body'—the hollow left in the surface of the ovary after the egg is released—that the German gynecologist Ludwig Fraenkel (following a suggestion of his teacher Gustav Born) had shown in 1903 to constitute a hormone-producing ductless gland. In 1931, in a remarkable book, *Die hormonale STERILISIERUNG des weiblichen Organismus*, of less than 15,000 words that hardly anyone now seems to have read, Haberlandt outlined in uncanny detail the contraceptive revolution of some 30 years later. He pointed out that oral administration, which he actually demonstrated in mice, would be the method of choice as well as the necessity for periodic withdrawal from the hormone to allow menses to occur. He called for the use of such contraception on clinical and eugenic grounds, arguing that it would enable parents to have the desired number of healthy children. Objections by people like the sexologist van de Velde that too many women would take advantage of hormonal contraception was dismissed by Haberlandt with the argument that such preparations would require a physician's prescription and would not be made available over the counter. He ended his manifesto with a visionary claim:

> Unquestionably, practical application of the temporary hormonal sterilization in women would markedly contribute to the ideal in human society already enunciated a generation earlier by Sigmund Freud (1898).'Theoretically, one of the greatest triumphs of mankind would be the elevation of procreation into a voluntary and deliberate act.'

Haberlandt did not limit his publications to the scientific literature. He also published in the popular press and gave interviews that led to huge newspaper headlines like 'My aim: fewer but fully desired children!' (in the 20 January 1927 issue of the *Acht Uhr Abendblatt, Berlin*) complete with commentary by the now-familiar chorus of physicians, lawyers and theologians. His obsession with the therapeutic potential of corpus luteum extracts was so well known that his students hung a banner by his home with the couplet, '*Verdirb nicht Deines Vaters Ruhm mit Deinem Corpus Luteum*' [Don't mar your father's renown with your corpus luteum]. But Haberlandt was not content with the visionary's role only. He contacted several pharmaceutical companies in an attempt to obtain consistently active and non-toxic corpus luteum and placental extracts for human clinical experiments. In his 1931 book, he finally reported success in the following words:

> I have been in contact for over three years with the therapeutic firm Gideon Richter in Budapest [to this day, a company active

> in the steroid field] and it is likely that in the near future a suitable 'sterilizing preparation' under the name 'Infecundin' will be available for systemic administration in clinical experiments as I had already announced in Vienna at the 4th Congress [September 1930] of the World League for Sexual Reform.

He confirmed that experiments in mice with orally-administered '*Infecundin*' had demonstrated temporary infertility without toxic reactions, 'since only in this manner does the new method have any chance for clinical success.' A year later, the 47-year old Haberlandt died, but the name 'Infecundin' survived. In 1966, it became the trade name of the first oral contraceptive produced in Hungary by the very same company Haberlandt had contacted 40 years earlier.

Within two years of his death, pure progesterone was isolated in no less than four laboratories in Germany, the US and Switzerland; its chemical structure established by Karl Slotta (who was, eventually, a Hitler refugee settling in Brazil); and its synthesis from the soyasterol stigmasterol accomplished. Had Haberlandt lived, there is no question that he would have pursued his dream of temporary hormonal sterilization in humans without resorting to glandular extracts. But even with pure progesterone, he could have shown only that ovulation can be inhibited by injection as the appropriately named American investigator A.W. Makepeace demonstrated in 1937 in rabbits and E.W. Dempsey in guinea pigs. Progesterone, being synthesized within the body, lacks the chemical features that would enable it to withstand the rigors of oral ingestion—the manifold enzymes and acids that stand between our bloodstreams and the world around us. For that, he would have needed another steroid—not naturally occurring, but waiting to be synthesized—and that took a further 20 years. Thus, nothing more happened, and Haberlandt's work fell into such oblivion that the next biologist to take it up, Gregory Pincus (who clearly should have known better), did not even feel obligated to cite Haberlandt among the 1459 references in his own *opus magnum, The Control of Fertility* (1965). Nor for that matter did Pincus's clinical collaborator, John Rock, whose book *The Time has come* (1963) quotes Makepeace's work but none of Haberlandt's pioneering earlier research. Yet if there ever was a grandfather of the Pill, Ludwig Haberlandt above all others deserves that honor.

III

A rather different addition to the Pill's early genealogy was the name of Russell Marker, a research professor at Pennsylvania State College in the late 1930s and early 1940s. Perhaps what attracted the journalists and TV filmmakers to add Marker to the list of fathers of the Pill was Marker's status as a maverick. Lacking the formal union card of a PhD, Marker's true rank as one of the giants of steroid chemistry was recognized only two decades after his sudden and total withdrawal from chemistry while still in his 40s. But despite his genuine claim to greatness in the larger field of steroids, in terms of oral contraception in particular I would rate him a shirt-tail relative.

Which is not to say that Marker wasn't important, albeit indirectly, for making the raw materials of contraceptive research more readily available. Until the mid-1940s, virtually all clinically useful progesterone was prepared in one way or another from soyasterols or cholesterol in processes that required conversion of such sterols into intermediate products prior to final transformation into progesterone. These intermediate steps represented a bottleneck, as only limited quantities of the necessary substances could be generated at one time. Not surprisingly, this poor yield kept the price of progesterone very high (approximately US$80/gram in the early 1940s). All this changed dramatically when Marker revolutionized the chemical production of progesterone. Within a few years, as a result of his process, the cost of progesterone dropped sufficiently that it became inexpensive enough to be used as the starting material for the synthesis of other steroids (for instance cortisone), rather than just serving as a clinically useful drug for menstrual disorders. But what, precisely, was the nature of Marker's discovery?

In the late 1930s and early 1940s, Marker conducted research on a group of steroids called sapogenins, compounds of plant origin. They got their name because, in their naturally-occurring form (where they are linked to sugars in compounds called saponins) they form soapy lathers in water. Natives of Mexico and Central America had long used them for doing laundry and to daze or kill fish. Marker concentrated on the chemistry of a member of this group called diosgenin, which was present in certain types of inedible yams (*Dioscorea* species) growing wild in Mexico. He succeeded in developing a five-step, high-yield conversion of diosgenin into progesterone. All kinds of apocryphal stories have been written about Marker's departure from Pennsylvania State College during World War II and his move to Mexico, many embellished with mysterious disappearances into the Mexican jungle, newspaper-wrapped parcels containing the equivalent of the world's supply of progesterone, and the like. But in 1979 (approaching the age of 80), he visited me at Stanford University and permitted a taped interview. Despite the roughness of the transcription that follows, I give it here in raw form, because it catches something of the character of the man.

Djerassi: *This is October 3, 1979, and I am finally having the meeting with the great Russell Marker, who can tell me what really happened in Mexico. Just tell me that one part over again—you collected about 10 tons of Dioscorea in Mexico . . .*

Marker: *After I was convinced that Parke-Davis* [the American pharmaceutical company that supported his chemical research at Pennsylvania State College, but refused to consider the industrial applications of that work] *would not go into it, I tried other companies to get support. For instance, I tried Merck and they said that since Parke-Davis turned me down they could not go into it . . . Then I decided that I was going to go into it myself and I withdrew from the bank about half of my meager savings and went to Mexico, where I collected 9 or 10 tons of root from the natives that found the original two plants for me. I collected that between Cordova and Orizaba, near Fortin. The man that had collected the original had . . . a little store and a small coffee-drying place right across the street. We collected material and he chopped it like potato chips and dried it in the sun, and I took it up to Mexico City and had it ground up. I found a man that had some crude extractors there; he extracted it with alcohol and evaporated it down to a syrup. And that I took back to the United States to a* *friend of mine who had a laboratory. I made arrangements with him that if he would do the rest of the financing and let me use his laboratory, I would give him one-third of the progesterone that we got. I told him that I expected a little over 2 kilos. But we ended up with having a little over 3 kilos and he took a kilo of it. At that time he was getting $80 a gram for it.*

Djerassi: *But how did you carry out the degradation of diosgenin to pseudodiosgenin* [the first step in Marker's published chemical procedure], *which, after all, was really an autoclave reaction?*

Marker: *He had a metal autoclave.*

Djerassi: *On what scale did you do that first? How much diosgenin did you degrade?*

Marker: *About 2 kilos at a time.*

Djerassi: *Had you ever done it on that scale at Penn State before?*

Marker: *Yes.*

Djerassi: *How did you meet Somlo?* [An ex-Hungarian lawyer/businessman and former principal owner of the small Mexican pharmaceutical firm Laboratorios Hormona, who then became one of the founders of Syntex.]

Marker: *I went to several people in Mexico with the hope that someone would be interested. I went to the telephone directory while I was staying at the Hotel Geneve . . .*

Djerassi: . . . *and you looked under 'hormona'?*

Marker: *No, I looked under 'Laboratorios' and I found Laboratorios Hormona and I thought they must be interested in making hormones. So I took a taxi and went out and Lehmann* [scientific director and minority stockholder of Laboratorios

Hormona] *was there. Lehmann looked at me—apparently he thought I was crazy or something when I first went in and then he excused himself and went out and when he came back he said, 'Oh, you are the Marker that has published these papers?'* [Marker had published all of his work earlier in the Journal of the American Chemical Society]. *He said it sort of rang a bell that, 'I have seen your name some place.'*

Djerassi: *He took you seriously then?*

Marker: *He took me seriously, and he wanted to know if I would be in town for a few days. He said that Dr Somlo, who owned the company, was in New York. So he called Dr Somlo and told him to come back immediately. And Dr Somlo came back the next day or so, and I had a talk with Dr Somlo; they wrote out a small contract that we would start the production and start a new company as soon as I would be available, and things like that. I told them that I had some research that I wanted to finish up before I came to Mexico and that they would finance it.*

Djerassi: *And did they know that you had made some progesterone in the States before?*

Marker: *No, I didn't tell them anything about it. So several months later I came back to Mexico, and as I was leaving I told Lehmann that I had made several kilos of progesterone in the States. He was greatly surprised at that and he wanted to know what I had done with it. I told him that I still had it in my possession and when I got back to the States, I got a phone call from Somlo; he wanted to know if I still had that progesterone. I told him that I did. He said, 'Meet me in New York in a few days,' and I told him I would. So he said, 'We will set up a company in Mexico for 500,000 pesos,'* *which was a little over $100,000 in those days—pesos were worth roughly 21 cents, and he said, 'We make a deal that you are going to have 40 per cent of the stock but you don't have any money to pay for it.' He said, 'Give me the progesterone and we will start selling the progesterone in this company.' So I made a deal with him, and he said we would take out the first $40,000, which I owed for the 40 per cent of the company that we were going to form. It didn't have a name at that time. And the rest of the money that we get for this—see, we were selling for $80 a gram at that time —we'll put into the company's profits and we will split the profits then so that I get 40 per cent, Lehmann 8 per cent, and he 52 per cent.*

Djerassi: *Who thought of the name Syntex? Was that Somlo or you?*

Marker: *Somlo came in one day when we were about ready to start the company and he said that he had a name for the company. 'Synthesis' he was going to call it. I asked him, since we were down in Mexico, why not have something to indicate that it was in Mexico. He said, 'All right, Syntex.' That's how it started.*

Djerassi: *Where did you set up a lab there, at Laguna Mayran?* [The name of the street in Mexico City where Laboratorios Hormona was located.]

Marker: *At Laguna Mayran—it was then the old Hormona building. See, then there was a vacant lot on a corner adjacent to Hormona and during the year they put up some laboratories there for me to work in and a place for the extractors. Well, after a year's time I went to Somlo. In the meantime, I spent all my money and my wife was in Mexico and I had to send her back because of being* *short of money, and it was cheaper for her to live in State College* [the location of Pennsylvania State College as the university was then named] *than in Mexico—even at that time. He would give me enough money to live on. From time to time I would go to him and say, 'I am short on money, I need some money to pay my hotel bills and send to my wife.' He'd give me $1,000 or something like that, you see, until I would come to him the next time.*

Djerassi: *You mean, he didn't even pay you a salary?*

Marker: *No, no salary, nothing.*

Djerassi: *Why did you do that?*

Marker: *The agreement that we had—we were going to split the profits, so at the end of the year, well it was probably in February or March, I went to him and I asked him about the profits because I knew there were substantial profits. I had made at least 30 kilos of progesterone, which I turned over to him, and some of it was going to Argentina and was reshipped to Germany during the war. That was another thing I objected to, that the product was being shipped to Germany; at least 2 kilos of it that I know happened that way. Well, I asked him about the profits because the progesterone at that time was still selling for $25 to $30 a gram; and he had reduced the price somewhat, and making some 30 kilos—well, you would have a profit of maybe half a million dollars or something like that. He said, 'What profits?' I said, 'The profits we made on Syntex.' And I asked if I could see the books. And he said, 'No, you wouldn't understand them anyway.' I told him I would get someone who speaks Spanish to look over the books. He said, 'I refuse to let you see them because you couldn't interpret them.' Finally he got pretty mad at me and he said, 'There is no profit at* *all.' I asked him where the profits were. He said, 'I took them as salary and you can't do anything about it.' So I decided to leave.*

When Marker checked the typed transcript of our taped interview he added the following paragraph, which describes his industrial activities in Mexico after his departure from Syntex and just before he withdrew totally in the late 1940s from chemistry:

When I left Syntex in May 1945, I formed a company known as Botanicamex in Texcoco and produced progesterone there until about March 1946, when production was moved to Mexico City with Gideon Richter [the Hungarian company that Haberlandt had contacted nearly two decades earlier], *who formed a company known as Hormosynth, later changed to Diosynth. While at Texcoco, I produced about 30 kilos of progesterone, approximately the same amount as I produced during my stay with Syntex.*

The interview ended on a poignant note. A driver had come to take Marker from my Stanford office to the San Francisco airport from where he was supposed to fly to Mexico City for a brief visit. He asked for the men's toilet and as I led him there, he suddenly turned to me, 'Tell me, where am I? What am I doing here? What did we talk about?' Fearing that he had suffered a sudden memory loss, I reached into his jacket and drew out his plane ticket. After explaining gently where he was, I gave the ticket to the driver, asking him not to just drop off Marker at the airport, but rather see to it that he boarded the plane. A few weeks later, I learned that Marker had suffered a mild heart attack and had been taken off the plane in Texas to a hospital.

I had conducted my taped interview in the presence of my colleague, Harry Mosher, now Emeritus Professor of Chemistry at Stanford, who had been a graduate student at Penn State and had worked in the same laboratory as Marker. After Marker departed, Mosher turned to me. 'There are other versions as well.' And sometime later, after I had published an autobiography in which Marker was mentioned, I received a letter from a reader, an American chemist now living in Israel, who had also worked in the same laboratory at Penn State. It was a startlingly bitter letter, citing evidence to support his view that Marker had been virulently anti-Semitic around the outbreak of World War II, a claim that was subsequently confirmed by another independent witness from the early 1940s.

In terms of the chemical history of the Pill (though not in the eyes of a refugee from Hitler, like myself), Marker's purported anti-Semitism would appear to be irrelevant. But there is an intriguing aspect to Marker's alleged prejudice. His PhD supervisor at the University of Maryland was Morris Kharasch, one of the very first Jewish professors in the WASP (White Anglo-Saxon Protestant)-dominated university chemistry faculties of pre-World War II America. For reasons that have never been completely clarified, Marker never finished his doctorate at Maryland, but, following a short industrial stint at Ethyl Corporation, he spent six years at the Rockefeller Institute working under another well-known Jewish chemist, Phoebus Levene. That collaboration broke off so bitterly that Marker—in a last gesture of defiance to a superior—removed the labels of virtually all the research samples that he left behind when he departed for his new position at Pennsylvania State College.

Among chemists, only burning one's laboratory notebooks would be a worse act of vandalism.

Readers may well ask whether I am just spinning out another thread of the Marker legend that so many journalists have woven. In this case, the evidence is more direct; that interview in Palo Alto was not the first time our paths had crossed. Some 20 years after Marker's stint in Levene's laboratory, I contacted his successor at Rockefeller, Alexander Rothen. I had no personal interest in Marker at the time; I was after a purely scientific quarry, a collection of optically active hydrocarbons that Marker had synthesized while working with Levene in the period 1928–1934. In response to my request, I was presented with boxes of glass vials containing the precious samples; imagine my feelings on discovering that most were unlabeled. Chemists, of course, are clever at uncovering the identity of unknown compounds—an activity known as 'structure elucidation'—and I was rather good at it. But it was such an annoying and unnecessary loss of time—a whole lab exercise to go through before I could even begin the work that had prompted my request—that I always wondered what had caused Marker to do it. Did antipathy toward Jews have anything to do with it? Interestingly, the Jewish connection did not end at Rockefeller. Marker's two partners during the founding of Syntex and the subsequent bitter rupture were Jewish emigrants from Europe.

The only subsequent time I met Marker was in 1984, when the annual Russell Marker Lectures in the Chemical Sciences at Pennsylvania State University were inaugurated. He requested that I present the first series and I used that opportunity to pay homage to him by paraphrasing 'Marker-at-Stanford' in front of Marker. Physically and mentally, he was in fine fettle and I was pleased that he wanted me to initiate the annual event established in his honor. Yet I wondered whether he knew that I was Jewish.

Still, Marker's enduring contribution to contemporary chemistry, and indirectly also to pharmacology and medicine, lies in his discovery that steroid hormones could be synthesized from a naturally occurring and cheap plant source—research that I considered so important that on one occasion I nominated Marker for a Nobel Prize. To appreciate the significance of Marker's work it is necessary to understand the difference between 'total' and 'partial' synthesis. To a chemist, 'total synthesis' is making a molecule from scratch—essentially from air, carbon sources (such as coal or petroleum), water and other elementary substances—not unlike building a house from clay and timber, iron and sand. 'Partial synthesis,' on the other hand, involves starting with an advanced structure—say a barn—and then converting it into a habitable house with plumbing and central heating. In Marker's synthesis, the building to be constructed was a 'steroid'—a term that one hears often enough in contemporary life, but rarely (outside of organic chemistry classes) hears defined. This is probably because the definition is not simple; I shall attempt to define it all the same, in the spirit of Einstein's famous dictum: 'We should make things as simple as possible but not simpler.'

The word Steroid, meaning 'like a sterol,' is derived from the Greek. Sterols, in turn, are solid alcohols (Gr. *stereos*, solid + ol) that occur widely in plants and animals—the best known being cholesterol, the most abundant sterol in humans and other ver-

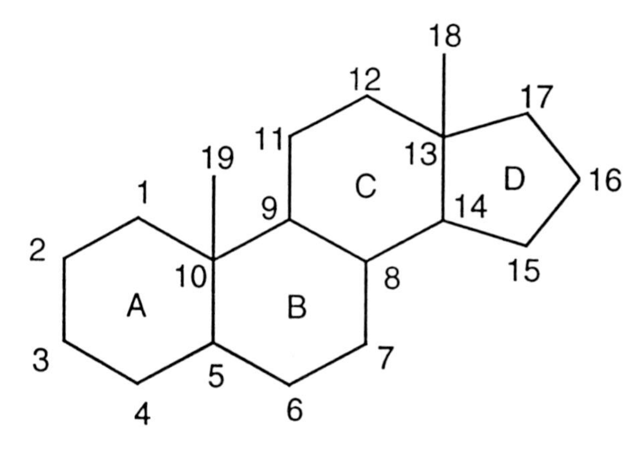

Figure 2.1 Shorthand notation of steroid skeleton.

tebrates. All steroids (and all sterols) are based on a chemical skeleton that consists of carbon and hydrogen atoms arranged in four fused rings (see Fig. 2.1) and known generically by the forbidding name perhydrocyclopentanophenanthrene. Steroid chemists communicate among themselves not in such polysyllabic jawbreakers, but in diagrams such as Fig. 2.1. They simplify life for themselves still more by dropping the symbols for carbon (C) and hydrogen (H). This shorthand is meaningful (to a steroid chemist) if you also assume the following: all angles are occupied by carbon atoms; all carbons are connected four ways to other atoms; any unaccounted-for connections are taken up by hydrogens. With these conventions in mind, you can read Fig. 2.1 as showing three rings (A, B, C) each with six carbons and one ring (D) with five. Some of the carbons have three unmarked hydrogens connected to them, others two, some one, and two (numbered 10 and 13) have none, all their bonds being occupied by connections to other carbons. Together, these fused rings define a steroid. Note that atoms 18 and 19 are not part of a ring but attached as methyl groups. (The full chemical notation of methyl is CH_3, but for shorthand purposes, it is simply written as a vertical straight line.)

Thousands upon thousands of synthetic, and many hundreds of natural, compounds are based on this fundamental steroid skeleton made up of carbon and hydrogen atoms as depicted in Fig. 2.1, all differing minutely in chemical structure by the attachment of some additional atoms (usually oxygen) at various locations, most commonly at positions 3 and 17. The variations, however minute, produce dramatically different biological results. Many of the most important biologically active molecules in nature represent slight variations on the steroid skeleton: the male and female sex hormones, bile acids, cholesterol, vitamin D, the cardiac-active constituents of digitalis, the adrenal cortical hormones (related to cortisone and usually referred to generically as corticosteroids) and many plant-derived products. The wide-ranging biological activity of steroids—for instance, the fact that one is responsible for male and another for female secondary sexual characteristics—is, in part, associated with the introduction of a third atom, oxygen (O), to special positions of the steroid skeleton.

When Marker started his research in the 1930s, the total synthesis of steroids had not yet been accomplished. All steroid hormones then known (e.g. progesterone and testosterone) were produced only by 'partial' synthesis from naturally occurring steroid precursors, mostly from cholesterol and bile acids—both of animal origin. By pursuing synthesis from such a starting point, chemists were simply imitating nature, which uses cholesterol as the starting material for the synthesis of steroids in the body. Although kilogram quantities of steroid hormones such as progesterone were produced by some drug companies, the methods were unwieldy. Marker's contribution was to show that compounds derived from his sapogenins could be transformed much more cheaply into the desired steroid 'house' (e.g. progesterone) than could a more cumbersome 'barn' such as cholesterol.

As he recounted in his interview with me, after he was unable to convince any American pharmaceutical firm of the commercial potential of diosgenin, Marker formed in Mexico in 1944 a small company, Syntex, in partnership with two European immigrants, Emeric Somlo and Federico Lehmann. A few months later, Syntex started to sell to other pharmaceutical companies (but not to the general public) pure, crystalline progesterone prepared from diosgenin by partial synthesis in five chemical steps. Within a year the partners had a disagreement, and Marker left the company, taking his technical expertise with him. For a time, Syntex was without its most profitable product.

Early in his academic career, however, while still at Pennsylvania State College, Marker had published a description of his chemical processes in the *Journal of the American Chemical Society*. Since no one had taken out patents in Mexico for his discoveries, the commercial production of progesterone from diosgenin was up for grabs in that country. Somlo and Lehmann, looking for another chemist who could re-establish the manufacture of progesterone from diosgenin at Syntex, recruited Dr George Rosenkranz from Havana. A Hungarian like Somlo, Rosenkranz had emigrated to Cuba a few years earlier from Switzerland, where he had received his doctorate under the Nobel laureate Leopold Ruzicka (one of the giants of early steroid chemistry) and was already familiar with Marker's publications. Within two years, Rosenkranz had re-instituted the large-scale manufacture of progesterone from diosgenin. Even more important, he had achieved the large-scale synthesis, from those same Mexican yams, of the commercially more valuable male sex hormone testosterone. Both syntheses were so much cheaper than the methods used by the European pharmaceutical companies then dominating the steroid hormone field—such as CIBA in Switzerland, Schering in Germany, and Organon in Holland—that in a short while tiny Syntex broke the international hormone cartel. As a result, prices fell, and these hormones became much more available. During the late 1940s, Syntex served as bulk supplier to pharmaceutical companies throughout the world, but few people outside these firms even knew of the existence of this small chemical manufacturing operation in Mexico City, which was soon to revolutionize steroid chemistry and the steroid industry all over the world. By the late 1950s, over half the world's supply of steroid hormones originated from Mexico, where in the meanwhile other American and European pharmaceutical companies had also started to establish manufacturing subsidiaries based on the commercial exploitation of diosgenin.

So why do I relegate Marker at best to the role of a distant maternal grand-uncle in the genealogy of the Pill? Because the availability of progesterone on an industrial scale did not contribute in any way to the development of oral contraceptives. Certainly a source of pure progesterone was necessary, and in sufficient quantities to investigate its therapeutic use. But the need of progesterone for contraception had been Haberlandt's insight, not Marker's, and the provision of pure progesterone had already been achieved in the 1930s, long before Marker and shortly after Haberlandt's death, through the work of German scientists like E. Fernholz and Adolf Butenandt, who won a Nobel Prize in 1939 for his steroid research. But nothing had happened, not because progesterone was expensive, or in short supply, but simply because it was not sufficiently active by oral ingestion to be taken as a pill. By the time Marker appeared on the scene with a new and better synthesis of progesterone, the steroid contraceptive boat had already sunk. It took two other maternal uncles and a mother to launch a new one, and a father to make it seaworthy.

Without Marker, of course, there would not have been any Syntex—the physical site in Mexico where the maternal role was played out. But there is no question that some other mother would have appeared elsewhere soon thereafter, because in the chemical sense, two older maternal relatives (Maximilian Ehrenstein and Hans Herloff Inhoffen), often completely ignored, had inadvertently pointed the way for a chemical mother to produce the requisite ripe egg—in other words, the creation of a synthetic steroid mimicking the biological role of progesterone, but being active by mouth. Before describing that 'maternal' process, a brief detour from oral contraceptives to cortisone is warranted because, while cortisone has no biological connection with the Pill, cortisone was nevertheless the molecule that brought me to the Pill. Furthermore, the truly important contribution of Marker's new progesterone manufacturing process is most clearly illuminated in the field of corticosteroids.

IV

By the fall of 1945, as a 22-year-old, newly-naturalized American citizen with a PhD degree from the University of Wisconsin and a wife, I returned to CIBA (the pharmaceutical firm in New Jersey where I had worked for one year after graduation from Kenyon College) for another four years, to resume work on antihistamines and other medicinal compounds. Since we were permitted to spend about 20 per cent of our time on independent research, I was able to continue on the side my ever increasing interest in steroids from my graduate school days. In spite of the industrial setting, I managed to publish a fair amount on my own in the scientific literature. I considered such publications crucial for eventual entry into academia at some advanced level.

This plan turned out to be somewhat naîve. At that time the path from industry to a university position was mostly a one-way street—in the wrong direction. A few chemists had shown that it was not impossible, but a great many more had demonstrated that it was extremely difficult. And I, at the ripe age of twenty-five and a half, with nearly five years of industrial experience behind me, had concluded that I was ready for an academic career. Even now, after almost five decades in academia, I am still not certain what fueled this ambition, other than the conventional snobbery picked up in graduate school that the academic ladder leads to nobler intellectual heights. But since I had absolutely no luck—very few interviews and no offers—I searched for chemical challenges on my own industrial turf.

The late 1940s were exciting days in steroid chemistry, especially since the anti-arthritic properties of cortisone had just been discovered and promptly recognized by a Nobel Prize to Reichstein, Kendall, and Hench. I was anxious to work on an improved synthesis of cortisone at CIBA, but, as most of that work was then conducted at the firm's headquarters in Switzerland, this proved impossible.

One day in the spring of 1949, I received an unsolicited employment offer from Syntex, a company I had never heard of. Although the position, as associate director of chemical research, seemed tempting to me, the location of Syntex in the professional backwater of Mexico made the offer seem ludicrous. Fortunately, I am a tourist at heart; when I heard the invitation, 'Come and visit us in Mexico City, all expenses paid,' I went. And as a bonus decided to include a visit to Havana in my itinerary.

George Rosenkranz, then technical director of Syntex and barely past 30, impressed me enormously as a sophisticated steroid chemist; he also charmed me personally. Rosenkranz showed me rather crude laboratories, but he promised lots of laboratory assistants and substantial research autonomy to devise a practical synthesis of cortisone and to pursue other aspects of steroid chemistry that might interest me. Furthermore, even though the labs were primitive (I still recall my amusement on observing a reaction vessel shaking in the sunshine of the open patio on Calle Laguna Mayran 413), Syntex could boast of some advanced equipment such as an infrared spectrometer at a time when neither CIBA nor my Alma Mater, the University of Wisconsin, had such an instrument, which proved to be enormously useful for steroid research.

I arrived at Syntex in the late autumn of 1949, just around my twenty-sixth birthday. I have never regretted that decision, even though at that time my American colleagues considered me mad to move to a country that, although famous for mariachi music, bull fights, and pre-Columbian ruins, had only generated the barest of blips on the radar screen of international chemical journals. These blips had consisted primarily of some long articles on steroidal sapogenins by Russell E. Marker that, although identified as originating from the Hotel Geneve, Mexico City, were (correctly) interpreted as the swan song of a maverick gringo chemist who was in the process of retiring permanently from the chemical scene.

Yet I was convinced that the best route to the academic job still eluding me was to establish a reputation in the scientific literature. I felt intuitively that Mexico was the right place for me. Syntex had the same objective I did: to establish a reputation. Our common goal—the 'partial' synthesis of cortisone from a plant raw material—was one of the hottest scientific topics in organic chemistry at that time. I was young and willing to gamble on a few years in Mexico—partly because living in another country and learning another language appealed to me, but also because I thought that any scientific achievement from a laboratory in Mexico was likely, on publication, to make a much bigger impression on academia than one coming from the usual elite laboratories in North America or Europe.

Consequently, I really had only one requirement before I accepted the Syntex offer, and that was to publish any scientific discoveries promptly in the chemical journals. Syntex agreed to this and stuck to its bargain. From my previous industrial experience, I fully understood that discoveries have to be patented by the firm in whose laboratory the work is performed before they are written up for publication. But instead of having patent attorneys deciding whether and when to publish, at Syntex Rosenkranz and I called the shots—extraordinary for a pharmaceutical company. As a result of this policy, during my first two years at Syntex we published more rapidly in the chemical literature than did any other pharmaceutical company, or even many university laboratories. Long before Syntex sold drugs under its own name to the medical profession, its international scientific reputation in chemistry was well established. Ten years after my temporary move to Mexico, when Professor Louis F. Fieser of Harvard analyzed in 1959 the references in the latest edition of his text *Steroids*, the recognized bible of steroid research, he found that no laboratory in the world—academic or industrial—had published as much in the steroid field as Syntex had in that time. Chemistry south of the Rio Grande had finally made the grade.

It was just as well to have so much incentive, because in undertaking a commercially viable synthesis of cortisone, we had taken on a daunting task. Until 1951, the only source of cortisone was through an extraordinarily complex process of 36 different chemical transformations starting from animal bile acids—a *tour de force* pioneered by Lewis Sarrett of Merck and Co. For many years, this had proved to be the longest and most complicated synthesis of any chemical on an industrial scale. Now that cortisone had emerged as a wonder drug, however, such a process looked less like a *tour de force* and more like a bottleneck. Developing an alternative partial synthesis from a plant raw material became one of the hottest scientific projects, with a number of powerful academic and industrial research groups competing to be first. At the outset, nobody even realized that a small research team in Mexico City had entered the race; when we completed our synthesis in June of 1951, ahead of everyone else and using Marker's stand-by diosgenin as the raw material, the resulting publicity was astounding.

Life magazine featured our team, most in immaculate white lab coats, grouped around a gleaming glass table and apparently mesmerized by an enormous yam root, which overwhelmed the molecular model of cortisone lying next to it. Rosenkranz held a test tube, filled almost to the brim with white crystals—the chemist's equivalent of the climber's flag on top of Mount Everest. For the photographer's benefit, the tube had been filled with ordinary table salt, because at that time we had synthesized only milligram quantities of cortisone. *Life*'s headline above the picture read, 'CORTISONE FROM GIANT YAM,' with the subsidiary headline 'Scientists with average age of 27 find big supply in Mexican root.' *Newsweek* had actually beaten *Life* by a week. Its headline read: 'SYNTHETIC CORTISONE—AND FROM YAMS,' followed by the statement that 'Unexpectedly the cortisone race was won by Syntex, Inc, of Mexico City.' But the quotation that probably made our entire team feel best appreciated—one last blast of the bellows on the flickering flame of our pride—appeared in the September issue of *Harper's Magazine*:

> As perhaps no other recent development, it [cortisone] . . . also underscores a point often overlooked in a big-money age. Big minds rather than big research budgets lead to big discoveries . . . Last but not least, it should be noted that the leader in the race was a chemical manufacturer in presumably backward Mexico.

Marker's connection with our cortisone work was, of course, indirect. He had already retired from chemistry and had left Syntex six years earlier. Neither Rosenkranz nor I had even met him. But our research on cortisone would never have been started, nor Syntex been founded, had it not been for Marker's earlier synthesis of progesterone from diosgenin. In the end, Marker's contribution toward a cheaper route to cortisone was not just seminal; it also proved to be indispensable for an unexpected reason.

Ironically, none of the cortisone triumphs from our laboratory and those of our competitors at Harvard and Merck that appeared in the August 1951 issue of the *Journal of the American Chemical Society* contributed to the treatment of a single arthritic patient because of the appearance of a newcomer, whose participation in the race was not even known. A few months after our publication, Syntex's management received an inquiry from the Upjohn Company of Kalamazoo, asking whether we could supply them with *ten tons* of progesterone. Since the world's entire annual production at that time was probably less than one-hundredth that amount, such a request seemed outlandish. No one in our group could conceive of a medical application of progesterone that would require tons of the stuff. We concluded that Upjohn was planning to use progesterone as a chemical intermediate rather than as a therapeutic hormone. Our conclusion proved correct when, a few weeks later, we learned through a patent issued to Upjohn in South Africa (where patents are granted much more rapidly than in the United States) that two of its scientists, Durey H. Peterson and Herbert C. Murray, had made a sensational discovery: fermentation of progesterone with certain microorganisms resulting in an one-step key transformation on the way to cortisone. What we chemists had accomplished laboriously through a series of complicated chemical conversions, Upjohn's microorganism with its own enzymes did in a single step in a few hours.

Still, our successful synthesis of cortisone from diosgenin had permanently placed Mexico on the scientific map of steroid research. It was, moreover, Upjohn's requirements for tons of progesterone—a quantity that at that time could be satisfied only from diosgenin through the Marker process—that started Syntex on the way to becoming a pharmaceutical heavyweight. So without Marker's synthesis, it is doubtful that any of Syntex's later successes would have been possible—for economic, if not for scientific reasons. This is why I assign Marker a branch, even if only a remote one, on the Pill's family tree, because our initial scientific success with cortisone, and the huge order from Upjohn, were followed almost immediately by our synthesis—a few months later, again in Mexico City—of an orally active progestational steroid. Which brings me back to a discussion of the maternal role in the birth of oral contraceptives, and the role of not one but two chemical uncles.

V

At the time that I became interested in the chemistry of progestational steroids, one of the dogmas of steroid chemistry was that almost any chemical alteration of the progesterone molecule would either diminish or destroy its biological activity. This belief is puzzling in light of the fact, well known at the time, that estrogenic steroid hormones, which occur naturally in a variety of forms, as well as synthetic chemicals not even based on the steroid skeleton, display marked estrogenic potency. In 1944, Maximilian Ehrenstein (another emigrant from Nazi Germany), then working at the University of Pennsylvania, published a paper that was mostly overlooked, but had made a deep impression on me while still a graduate student. By an extremely laborious series of steps, Ehrenstein had transformed the naturally-occurring steroid cardiac stimulant strophanthidin into a few milligrams of impure '19-norprogesterone.' To return to my earlier metaphor, Ehrenstein had transformed a very elaborate mansion (strophanthidin) into a funky little vacation house. While he had obtained only enough material for biological testing in two rabbits, in one of them his compound had displayed higher progestational activity than the parent hormone. A positive test in one animal out of two could, of course, have been just a fluke. What made Ehrenstein's results so unusual was what that '19-nor' in the compound's name signified. It meant that Ehrenstein had removed carbon atom No. 19 (between rings A and B of the steroid skeleton depicted in Fig. 2.1) from the most inaccessible site of the steroid molecule to replace it with a hydrogen atom. On paper—or in words—the change sounds trivial.

Given the state of the art of organic synthesis at the time, however, this was so difficult an operation that it had required several years for completion. Moreover, if the biological results were real, Ehrenstein's observation demolished the previous assumptions about the inviolability of the progesterone structure. But there was another problem: Ehrenstein's oily product was, as I indicated, impure: a mixture of at least three 'stereoisomers'—molecules that, while structurally identical, were, like mirror images, as alike—and fundamentally different—as your left hand and your right. In biochemistry, which often requires molecules to fit together like a hand in a glove, such a difference can be crucial. Which one of the components, if any, was responsible for the putative progestational activity? It took seven years for someone to come up with an answer. Our ability to do so led us almost straight to the Pill.

Part of my PhD thesis at the University of Wisconsin in the early 1940s had dealt with the partial synthesis of the then-inaccessible estrogenic hormones from the more readily available androgens, such as testosterone. For years, the estrogens were only available by isolation from the urine of pregnant women (and later of pregnant mares, the source of one of the more-frequently prescribed estrogen compositions in use today for hormone replacement therapy). In fact, the estrogenic hormones, such as estradiol and estrone, were the last steroid types to yield to partial synthesis, because no obvious precursor for them existed in nature. All of the other naturally-occurring steroids are based on the skeleton in Fig. 2.1; the estrogens, however, are based on the template shown in Fig. 2.2. Here, ring A has changed from the ordinary six-sided form to an 'aromatic' form,

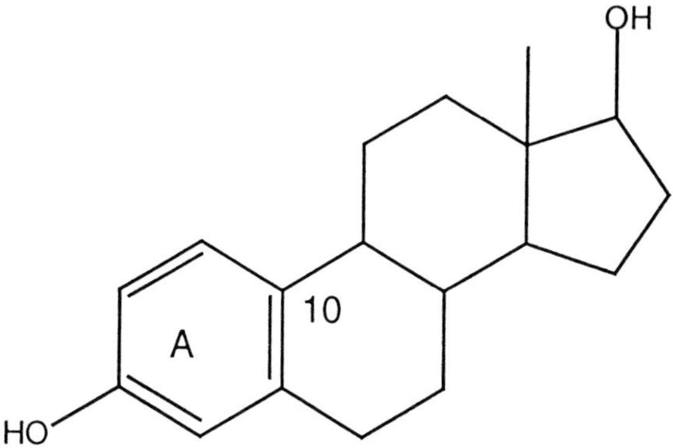

Figure 2.2 Estradiol (estrogenic hormone).

where half the carbon–carbon bonds are double. You don't need a PhD in organic chemistry to note the other difference between the estrogens and all other steroids. This is the absence of carbon atom 19 usually attached at position 10, which makes possible the doubling-up of carbon-carbon bonds in ring A. Chemically speaking, the only difference between testosterone (a conventional steroid with carbon atom 19) and the estrogens (aromatic steroids lacking C-19)—between men and women—is that one carbon, but what a difference it makes!

The partial synthesis of steroid hormones such as testosterone and progesterone could be extended to the estrogens if a process could be devised that would eliminate the key carbon atom No 19 and thus effect the 'aromatization' of ring A so typical of the estrogens. Hans H. Inhoffen, at Schering A.G. in Berlin, had demonstrated the practical feasibility of such a chemical conversion, but the work had been performed during World War II and experimental details were scant and had to be partly reconstructed. Syntex had started to use the Inhoffen process (which had not been patented in Mexico) for the production of modest quantities of estrone and estradiol. Upon assumption of my research position there, I suggested to Rosenkranz that Syntex examine another and potentially proprietary route to the estrogens directly from testosterone. In less than three months we succeeded in accomplishing this aim, which in chemical jargon would be described as the 'aromatization of ring A of conventional steroids.'

Our partial aromatization studies turned into the impetus that led us in a fairly straight path to the first synthesis of an oral contraceptive. From a technical standpoint, I felt that the time was ripe to follow up on Ehrenstein's lead of 1946. Using various chemical methods developed as part of our estrogen synthesis as well as methodology perfected by the Australian chemist, Arthur J. Birch (subsequently a long-term Syntex consultant), my Syntex colleagues and I prepared for the first time in 1951 pure, crystalline 19-norprogesterone (a steroid that like the estrogens lacked carbon atom 19) which, when assayed in rabbits at Endocrine Laboratories in Wisconsin, was found to be four to eight times as active as natural progesterone. In other words, Ehrenstein's observation with an oily mixture tested in one rabbit was more than confirmed: replacement of carbon atom 19 by one hydrogen had produced the most active progestational steroid known at that time.

With that lead in hand, we turned to another accidental discovery that had been made in 1939 in Germany, where chemists at Schering, again under the leadership of Inhoffen, found that if an acetylene group (a pair of carbon atoms connected by a triple bond) is added at position 17 of the male sex hormone testosterone, its biological activity is changed markedly: for unknown reasons this compound has weak progestational activity. Far more important, it was able to survive absorption through the digestive tract. On the reasonable assumption that removal of the 19-carbon atom increases progestational potency and addition of acetylene confers oral efficacy, Rosenkranz and I put both these observations together. On 15 October 1951, Luis Miramontes, a young Mexican chemist doing his undergraduate bachelor's thesis work at Syntex under my tutelage, completed the synthesis of the 19-nor analogue of Inhoffen's compound—that is, 19-nor-17α-ethynyltestosterone or, for short, 'norethindrone'—which turned out to be the first oral contraceptive to be synthesized. Lecture audiences are always intrigued when I display a slide showing the carefully dated and hand-written lab protocol of the very last step in that synthesis conducted by Miramontes, in which the elements of acetylene are added to impart oral activity. The notebook page, dated 15 October 1951, starts with the sentence, 'Se disuelve 1.0 g. de eter enol de 19-norandrostendiona en 25 cc. de Tolueno anhidro . . .'—in other words dissolving 1.0 g of the enol ether of 19-norandrostenedione in 25 cc of anhydrous toluene. In those pre-oncophobic years, when the carcinogenic properties of benzene were not yet recognized, the conventional medium of performing that acetylene reaction would have been a benzene solution. But we were working in 1.5 mile-high Mexico City, where the boiling point of benzene would have been too low. Hence we substituted it with the higher boiling toluene. This was the type of high-altitude chemical cookery that most gringos working in the fancy sea-level laboratories of Harvard or Merck never even had to consider.

We immediately submitted the compound to our favorite commercial testing laboratory in Wisconsin for biological evaluation and gloated happily when Dr Elva G. Shipley reported back that it was more active as an oral progestational hormone than any other steroid known at that time. In less than six months, we had accomplished our goal of synthesizing a superpotent, orally active progestational agent! Returning to my reproductive metaphor, I classify our synthesis of norethindrone as the release of the fertile egg waiting now to be fertilized. Once that is understood, the roles of Ehrenstein and Inhoffen as older maternal uncles become obvious. Since Ehrenstein lived in the States, our paths crossed in the 1950s at several scientific meetings. I know that he was pleased at our elevation of his original 19-norprogesterone work from a piece of chemical esoterica to one of seminal significance; we actually published a joint paper in 1958 to establish the nature of one of the components of his original 19-norprogesterone mixture. Inhoffen was different. We met only once at an international scientific congress. There, his comments seemed frosty, leaving the impression that my work as a graduate student at the University of Wisconsin had constituted an intrusion into his early work on the partial synthesis of estrogens. But in 1999, our paths crossed again, twice, though on his part posthumously. Early that year, I received the Inhoffen Medal at the Technical University of Braunschweig, but a more moving event occurred later that year in Graz. I had given a typical academic talk on the History of the Pill and had done so in German, which meant that it moved slower than it would have in English. When I realized that I would be running out of time, I decided to skip some slides. One of them was a picture of Inhoffen together with the father of Chinese steroid chemistry, Huang Minlon. The room was crowded and I had to cope with many questions before the audience broke up. Suddenly a tall, serious man, probably around 60 years old, approached me to ask quietly, 'Did you know Inhoffen and his work?' Before explaining what I had intended to say about Inhoffen, I produced the slide I had with me but had skipped during my talk. That's when I found out that I was speaking to Peter Inhoffen, a Catholic theologian and only son of Prof Inhoffen, from whom he had become completely estranged. I couldn't read his expression: was his question prompted by curiosity or by still smoldering filial pride?

That meeting finally prompted me to examine one aspect of Inhoffen's past that, like Marker's alleged anti-Semitism was not germane to his chemistry, but certainly was to my perception of him, and to my sense of my place in a vast network of researchers whose work depends so intimately on that of others. It was a rather obvious question for someone with my paranoia-inducing background as a Hitler refugee. Had he been a Nazi? My inquiries ultimately produced a letter written to me by one of Germany's most distinguished contemporary chemistry professors. I think it provides the best answer:

> I would say that Inhoffen was what in this country is known to be 'ein kleiner Nazi' [a little Nazi] . . . He became the first Rector of the Technical University of Braunschweig and since this was

> more or less an appointment by the British, I suspect that Inhoffen was more or less clean. He certainly was not an anti-Semite . . . His eternal problem—in practically all fields—was that he never was first. His competitors were not really better, but they were faster. Until his death, he could not define his role and position in the pyramid of competition, which is so characteristic of our profession. The old problem of internal and external view of yourself.

I wonder how he would have felt learning about his placement in my genealogy of the Pill? Pleased? Or scooped again?

VI

What follows now will seem to many to be just an overkill of dates. But since autobiographical reminiscences are all too frequently contaminated by automythology, a blemish the Pill's genealogy does not deserve, I resort to a scientist's typical compulsion about dates and publications to document the historical facts. The patent application for norethindrone was filed on 22 November 1951 (it is the first patent for a drug listed in the National Inventors Hall of Fame in Akron, Ohio), and I reported the details of our chemical synthesis, together with the substance's high oral progestational activity, at the April 1952 meeting of the American Chemical Society's Division of Medicinal Chemistry in Milwaukee. The abstract of this report under the names of Djerassi, Miramontes and Rosenkranz was published in March 1952, and the full article with complete experimental details appeared in 1954 in the *Journal of the American Chemical Society*. Non-scientist readers may well be irritated by such an avalanche of dates, but chronological precision is the baggage of scientists preoccupied with priority —a foible I would be stupid to hide.

A few weeks after having synthesized the substance and having received from Dr Shipley confirmation of its anticipated oral progestational activity, we sent it to various endocrinologists and clinicians: first to Roy Hertz at the National Cancer Institute in Bethesda, Maryland and to Alexander Lipschutz in Chile; later to Gregory Pincus at the Worcester Foundation in Shrewsbury, Massachusetts, to Robert Greenblatt in Georgia, and to Edward Tyler of the Los Angeles Planned Parenthood Center. It was Tyler who, in November 1954, presented the first clinical results of using norethindrone for the treatment of various menstrual disorders and fertility problems. All of these biological investigations can be equated to sperm that is surrounding the egg. But since this is a record of the history of the Pill, we need to address the source and origin of the particular sperm that led to the fertilization of our chemical egg and thus to the ultimate birth of an oral contraceptive.

Initially, we were not focusing on contraception when we developed an oral progestational compound. Our research was undertaken because at that time progesterone was used clinically for treatment of menstrual disorders, for certain conditions of infertility, and at a research level, for the treatment of cervical cancer in women by local administration of a high dose of the hormone. Such administration was extremely painful because it involved injecting a fairly concentrated oil solution of large amounts of progesterone into the cervix. What drove us was the desire to create a more powerful progestational compound that would be active orally. As it happened, the progesterone treatment of cervical cancer did not pan out, but the clinical use of our norethindrone (under the trade name Norlutin) for the treatment of menstrual disorders was approved by the FDA in 1957 and is one of its therapeutic indications to this day.

Each of the biologists mentioned above had his own area of expertise and interest in the field of progestational activity. Gregory Pincus and his colleague Min-Chueh Chang of the Worcester Foundation for Experimental Biology in Shrewsbury, Massachusetts, were focusing on how progesterone worked to inhibit ovulation (the mechanism behind Haberlandt's 'temporary hormonal sterilization' and Makepeace's confirmation). Among the many steroids tested in 1953 by the Worcester Foundation group for such activity two substances stood out: our norethindrone and another substance, norethynodrel, that had been synthesized by Frank Colton at G.D. Searle, a pharmaceutical company in the Chicago area. The chemical history of norethynodrel is worth telling, since it illustrates one of the less attractive features of scientific research: the drive for priority and recognition. In this instance, the stakes were higher than usual, since commercial considerations and financial returns quickly entered the equation.

For historical accuracy and appropriate credit, it is important to note that even though norethynodrel was synthesized well over a year following the publication of our successful synthesis of norethindrone, it was norethynodrel that first entered the market as an oral contraceptive. M.C. Chang had found norethindrone and norethynodrel to have been the two most promising candidates in his initial animal studies. But his boss,

Gregory Pincus, who was a consultant for Searle, selected the Searle compound for further work. Syntex, not having any biological laboratories or pharmaceutical marketing outlets at that time, licensed Parke-Davis & Co of Detroit to pursue the FDA registration and market the product in the United States. (That choice itself was ironic, since it was Parke-Davis that had sponsored the original research work of Marker's at Pennsylvania State College, but had then refused continued support in Mexico, thus leading to the foundation of Syntex). It was only after 1957, when both norethindrone and norethynodrel had entered the market (as drugs for non-contraceptive, gynecological purposes only), that the paths of the two companies diverged.

Searle deserves full credit for reaching the market first with an oral contraceptive, norethynodrel, under the trade name Enovid, but its repeated claim to have synthesized the substance independently and concurrently with Syntex's norethindrone constitutes a blatant misrepresentation of the facts. The record based entirely on published data is unambiguous. On 31 August 1953—well over one year after our first publication (March 1952) dealing with the synthesis of norethindrone, and 21 months after our own November 1951 patent filing date—Frank Colton of G.D. Searle & Company filed a patent application for the synthesis of a steroid that differed from norethindrone only in the position of one double bond connecting two carbon atoms (i.e. positions 5–10 rather than 4–5 in Fig. 2.1). In other words, Colton's compound was what chemists term an 'isomer' of ours—a compound of precisely the same atoms, differing only in their arrangement—in this case, in the most trivial of differences imaginable. Trivially different, because treatment of Colton's isomer, norethynodrel, with acid, or just human gastric juice, converts it to a large extent into Syntex's norethindrone—a conversion that among others was established by Gregory Pincus and his collaborators. Is synthesis of a patented compound in the stomach an infringement of a valid patent? (Interestingly, years later, a similar suit with a related oral contraceptive was pursued by Wyeth, Inc against Ortho Pharmaceuticals and initially resolved in favor of the plaintiff.) I urged that we push this issue to a legal resolution, but Parke-Davis, our American licensee, did not concur.

Searle was selling a very successful anti-motion-sickness drug, Dramamine, which contained Parke-Davis's antihistamine Benadryl. Given that the only FDA-approved use of our norethindrone in 1957 was the treatment of menstrual disorders and certain conditions of infertility, the issue seemed small potatoes to Parke-Davis, over which it was not worth fighting with a valued customer.

In the mid-1950s, Searle actively supported clinical trials of the contraceptive efficacy of norethynodrel. The work was conducted in Puerto Rico, under the direction of Pincus and especially John Rock, a clinical endocrinologist and gynecologist from Harvard. Around the same time in Mexico City and Los Angeles, Syntex sponsored contraceptive trials with norethindrone. But fearing a possible religious backlash, Parke-Davis suddenly chose not to pursue these results through the FDA approval process, and returned the contraceptive marketing license to Syntex. Alejandro Zaffaroni, Syntex's Executive Vice President, eventually negotiated a favorable marketing agreement with the Ortho Division of Johnson & Johnson, a company with a long-standing commitment to the birth-control field, but the shift to a new company meant a delay of nearly two years before Syntex's norethindrone received FDA approval as a contraceptive. By 1964, three companies—Ortho, Syntex, and Parke-Davis (having changed its mind after realizing that no Catholic-inspired boycott had developed)—were marketing 2.0 milligram doses of Syntex's norethindrone (or its acetate), which by then had become the most widely used active ingredient of the Pill.

Norethindrone's entrance into the market was nearly two years later than Searle's introduction of the first steroid-based contraceptive pill. There is no question that the company deserves kudos for marketing the product first—despite a possible consumer backlash by opponents of contraception. But given the extraordinary importance of these steroids, why does the Searle group to this day not disclose in the peer-reviewed literature any of the chemical research that led them to their pill? The only date supporting the claim for 'independent simultaneous discovery' is Searle's patent filing date of 31 August 1953, a date that sounds 'simultaneous' only without juxtaposition to the 21 November 1951 date of Syntex's invention of norethindrone.

Colton and other researchers from Searle had not otherwise been reluctant to publish their steroid research. Indeed, in 1957, they published an article in the *Journal of the American Chemical Society* about a new steroid anabolic, 17α-ethyl-19-nortestosterone (Nilevar), that was produced in one step from our 17α-ethynyl-19-nortestosterone (norethindrone). In their article, the Searle chemists cited quite properly our earlier publication, which predated theirs by over three years. Thus, their unaccountable shyness about the most significant product in their corporate history, norethynodrel, can only raise questions. Why, for instance, did Gregory Pincus, the person most responsible for persuading Searle to market norethynodrel, make not a single reference to any chemist (not even Frank Colton) in his 1965 *opus magnum, The Control of Fertility*? Why does his book make no mention of how the active ingredient of the Pill actually arrived in his laboratory? And even worse: why did Pincus drop Haberlandt's name into a black hole of anonymity?

Aside from not wanting to concede in the scientific literature Syntex's priority, I can think of only one other reason that Searle's chemical work in this field has never been seen in the bright light of a peer-reviewed journal; this hypothesis is pure speculation, and I confess I have not a shred of hard objective evidence to support it. But I did air it once in the published record of a symposium dealing with the chemical antecedents of the Pill held in New York City in 1992 by the American Chemical Society's Division of History of Chemistry, at the only time when Searle's Frank Colton and I appeared jointly at a scientific conference. My source was Leon Simon, a respected patent attorney practicing in Washington, DC since 1945, who specialized in the steroid drug field and served from the late 1940s through the middle 1960s as Syntex's independent patent counsel. (In 1965, after the Syntex research and corporate headquarters had moved to the Stanford Industrial Park, Simon followed the company to head Syntex's in-house patent department.) Several years before his death in 1975, he confided to me his own and presumably unprovable supposition of the genesis of Searle's August 1953 patent application concerning norethynodrel.

According to Simon, in January 1952 Dr Emeric Somlo, then the owner of Syntex, had some negotiations with the Searle family about the possible purchase of his company. To facilitate Searle's examination of Syntex's assets, Somlo instructed Simon to permit the late Dr A.L. Raymond, Searle's research vice president, to inspect all of the then-pending Syntex patent applications. One of these was our Mexican patent application of 22 November 1951, which disclosed the structure, the progestational activity, and the technical specifics of the synthesis of norethindrone. What, if anything, Raymond did—consciously or subliminally—with this proprietary information after returning to the headquarters of G.D. Searle and Company in Skokie, Illinois, will never be known. Yet in 1957, Raymond and Colton published their only joint paper in the *Journal of the American Chemical Society* on several very close chemical relatives of Syntex's norethindrone, though never on norethynodrel.

My preoccupation with establishing unequivocally the priority—and thus the metaphoric maternal identity—is not just my admittedly strong competitive drive. I am realistic enough to acknowledge that it really does not make any difference to the world who does what first. But giving credit to Syntex as the corporate institution where it all first started is important to me (even though I severed all connections with that company in 1972), because institutional memories are so short. Syntex was the first and possibly the only significant example of important research in such a highly competitive field being conducted in a developing country. Both qualitatively and quantitatively, the research output of Syntex during the 1950s has never been matched in the steroid field; the pride and self-assurance it provided to a cadre of Mexican organic chemists, virtually all of them trained at Syntex, was moving to witness. Yet that company does not exist anymore, because in 1994 it was acquired by the Swiss pharmaceutical colossus Roche and promptly swallowed and digested. In that digestive process, the entire research division of Syntex in Mexico, which had just moved into new quarters in Cuernavaca, was closed and all research personnel dismissed. To me, the cold-bloodedness of this corporate amputation seems unforgivable: I know of no other pharmaceutical company in Mexico that has currently any significant research presence.

Syntex, as a company, and Mexico, as a country, deserve full credit as the institutional site for the first chemical synthesis of an oral contraceptive steroid—a statement that is not meant in any way to denigrate Searle's commitment to the contraceptive field and that company's successful drive to be the first on the market with a steroid oral contraceptive. Interestingly, Syntex-developed norethindrone is still a widely used active ingredient of oral contraceptives, whereas Searle's norethynodrel disappeared from the market many years ago, to be superseded by other 19-nor steroids, all of which are close chemical relatives of norethindrone. But there is a more charming end to this story. In the process of swallowing Syntex, Roche not only closed the Mexican research laboratories where norethindrone was first synthesized. Roche also decided to distance itself from any involvement in the contraceptive field, and promptly sold the entire Syntex oral contraceptive line, still based in its entirety on norethindrone. Who was the purchaser? None other than G.D. Searle—the company that went to heroic lengths to circumvent the Syntex patent on norethindrone and now had to pay good money to market it as its lead oral contraceptive long after the original patent had expired. What a closure to an historical circle!

VII

It should not be surprising that I, as a chemist, in terms of my reproductive metaphor that equates any synthetic drug to an egg, spent the bulk of this chapter examining the maternal lineage of the Pill. But just as it is clear that Ludwig Haberlandt merits definition as the paternal grandfather, Gregory Pincus—despite the uncertainties of paternity generally—deserves to be called a father of the Pill. The initial rabbit experiments by M.C. Chang in Pincus's laboratory clearly were the sperm that fertilized the chemical egg, and the subsequent implantation of the embryo and eventual fetal growth can largely, though not entirely, be ascribed to further experiments conducted in Pincus's laboratory. But Pincus was not only a prolific and highly experienced endocrinologist, he was also a charismatic entrepreneur. Many times, this latter quality is more difficult to find than mere scientific brilliance; it took entrepreneurship of Pincus's caliber to bring the steroids provided by the chemist to the stage where clinical trials of the Pill could be initiated and where John Rock, as leader of the clinical team, could assume the mantle of metaphoric obstetrician for the eventual birth of the Pill. While Rock's name is inexorably connected with that role, others, notably Celso-Ramon Garcia (the first professor of obstetrics/gynaecology at the University of Puerto Rico Medical School) and Edith Rice-Wray (Medical Director of the Puerto Rican Family Planning Association) contributed heavily to the planning and implementation of the first clinical trials in the San Juan area. Rice-Wray subsequently directed a Family Planning clinic in Mexico City where she continued her clinical studies, this time with Syntex's norethindrone.

Of the numerous talks and interviews that I have presented over the course of decades on the birth of the Pill, three stand out in my mind. Two of them were formal occasions directly associated with Pincus's memory: the Gregory Pincus Memorial Lecture and Award presented in 1982 on his home turf, the Worcester Foundation for Experimental Biology, and the last Gregory Pincus Memorial Lecture at the 50th and final Laurentian Hormone Conference in 1993. Ironically, this event was held in Puerto Rico, although these annual meetings, founded by Pincus, usually met in the Laurentian Mountains of Quebec. But the most relevant one to my story is an unusual session held Friday morning, 5 May 1978, in an old New England mansion on the outskirts of Boston, the headquarters of the American Academy of Arts and Sciences. The Academy was holding a closed two-day session on 'Historical Perspectives on the Scientific Study of Fertility.' The purpose of the meeting was to have a free-flowing dialogue among some of the key scientists who had been active in the field of fertility in the United States during the previous 40 years (therefore, it was not surprising that, as far as I could tell, at age 55, I was the youngest of that group) in order to collect a record that historians of science might draw on in the future.

The unedited transcript of that Friday morning session reads awfully: Nouns do not match verbs, tenses get mixed, punctuation is lost, and many words are misspelled or appear to be inaudible. Nevertheless, one gets a real flavor of excited human dialogue and interruptions, of hurt egos, of hitherto undisclosed vignettes. Here are two samples.

Hechter: *May I take a couple of minutes?*

Djerassi: *I haven't finished. I'd like to continue because I've only gotten to the first half of my story.*

Reed: *He can have my time. This is the first really fruitful . . . (inaudible)*

Greep: *This is history from the horse's mouth, and I think it's very good.*

Djerassi: *I misunderstood. Did you want me to continue?*

Greep: *Yes.*

The scientific co-chairman of the Boston Academy's May 1978 meeting was Roy O. Greep, a distinguished endocrinologist at Harvard, who had known personally most of the actors in this play. Another key participant was Oscar Hechter, who for many years had been senior scientist of the Worcester Foundation for Experimental Biology. Though not directly involved in the development of oral contraceptives, he had been an intimate collaborator of Gregory Pincus. James Reed of Rutgers University was a historian studying the birth control movement in America.

I felt that this was the one opportunity, years after Pincus's death, where I could find out why he had been so ungraciously selective in not acknowledging work of others that was crucial to the development of the Pill. John Rock, who had not behaved very differently, was in the room, but he had reached an age where it was not possible anymore for him to contribute to the dialogue. His was a silent, poignant presence. But Celso-Ramon Garcia, Rock's and Pincus's closest clinical colleague, was present, which led to the following exchange:

Garcia: *Basically, the monograph 'Control of Fertility' that Pincus wrote expresses in detail what his feelings were about who contributed to what.*

Djerassi: *Why did he not mention any chemists, do you happen to know that?*

Garcia: *He was a biologist, the same way as you are principally presenting your story as a chemist.*

Djerassi: *That's not true. That's why I submitted a paper here with biological references, including yours.*

Garcia: *Well, okay, but the fact is that principally you are a chemist and your major contribution has been that of a chemist.*

Djerassi: *But this would be like my describing the history of oral contraceptives without a single reference to Pincus or Rock or yourself!*

Chapter 3

Bitter Pills

I

When I first encountered the capital P in some news article, I visualized its author as a macho journalist of the late 1950s—sleeves rolled up, eyes squinting through the smoke of the cigarette dangling from the corner of his mouth, two index fingers producing a machine gun rattle on the Remington—who, while writing some pithy piece on oral contraceptives, decided to capitalize the word *pill*, and thus inadvertently converted this pedestrian generic term into a powerful four-letter epithet. Since then, the Pill has been described as everything from a woman's panacea or her poison to the cause for the social emasculation of men. In the 1970s, in preparation for my first book addressed to a general audience, *The Politics of Contraception*, I re-read Aldous Huxley's 1958 *Brave New World Revisited*—his magisterial reflections on his *Brave New World* of 1932. That's where I found a previously capitalized pill, framed within startlingly relevant words of wisdom:

> [The population problem] is becoming graver and more formidable with every passing year. It is against this grim biological background that all the political, economic, cultural and psychological dramas of our time are being played out . . . The problem of rapidly increasing numbers in relation to natural resources, to social stability and to the well-being of individuals—this is now the central problem of mankind; and it will remain the central problem certainly for another century, and perhaps for several centuries thereafter . . . Obviously we must, with all possible speed, reduce the birth rate to the point where it does not exceed the death rate. At the same time, we must, with all possible speed, increase food production, we must institute and implement a worldwide policy for conserving our soils and our forests, we must develop practical substitutes, preferably less dangerous and less rapidly exhaustible than uranium, for our present fuels . . . But all of this, needless to say, is almost infinitely easier said than done. The annual increase of numbers should be reduced. But how? . . . Most of us choose birth control—and immediately find ourselves confronted by a problem that is simultaneously a puzzle in physiology, pharmacology, sociology, psychology and even theology. 'The Pill' has not yet been invented.

For years, I was convinced that this was the first printed appearance of 'the Pill.' I was disabused of that notion by the latest edition of the *Oxford English Dictionary*. It quotes one C.H. Rolph writing in 1957 about 'the quest now going on for what laymen like myself insist on calling 'the Pill'.' But who was C.H. Rolph and in what context did he write these words? I had never heard of him, but anyone credited with the first modern usage of 'the Pill' certainly merited some recognition. I tracked him down at last in Britain's used book capital, Hay-on-Wye, where I found one of his memoirs, *Further Particulars.* I was pleased to note that someone with his wide-ranging interests was apparently the linguistic father of 'the Pill.' For a start, the centenary of his birth (as Cecil Hewitt) coincides with the fiftieth anniversary of the Pill. He started his career with the London City Police, from which he retired after 25 years at age 45 with the rank of Chief Inspector. He began to write while still on the police force, publishing under the pen name C.H. Rolph in periodicals ranging from the *Police Reporter* to the *New Statesman.* He pursued this second career full-time as a journalist, essayist, book author, BBC interviewer and more. His tastes—for music, literature, law and liberal social policy—were so diverse that his memoir reads like a mini 'Who's Who' of twentieth-century Britain. His observations on the Pill appeared in the introductory chapter of an anthology, *The Human Sum*, for which Hewitt alias Rolph had managed to assemble a stellar list of authors, including Julian Huxley and Bertrand Russell. This is what Rolph had to say:

> He [Dr A.S. Parkes of the Medical Research Council] gives a modestly exciting account of the quest now going on, in biological laboratories in various parts of the world, for what laymen like myself insist on calling 'the Pill'; and by this phrase, which, like all men of science, Dr Parkes would doubtless reject, I mean the simple and completely reliable contraceptive taken by the mouth. [In point of fact, Parkes in his chapter makes no mention whatsoever of then already on-going research with orally effective steroids, thus showing Rolph as an even more successful prophet]. This, it can hardly be doubted, will one day

> become available for the control of human fertility, universally, among the most backward as well as the most advanced communities in the human race; and its tremendous implications must, in the soberer thoughts of any person with social compassion, dwarf any other consideration that this book can provide.

I could not help but be struck that this coinage of 'the Pill' was prompted by A.S. Parkes, another author in Rolph's *The Human Sum*. In 1993, I had the privilege of delivering the first annual Parkes Memorial Lecture of the Society for the Study of Fertility at Cambridge University in honor of that distinguished British reproductive biologist. History is indeed circular!

Still, even if Huxley was second in print, he was certainly the most elegant proponent of the now accepted capitalized version of the Pill. Within three years of his pronouncement, the FDA approved the use of oral progestational steroids for contraceptive application and not long after, the Rolph/Hewitt-Huxleian usage was consecrated in both the American *Webster*'s and the *Oxford English Dictionary*: '*often cap*: an oral contraceptive—usu. used with *the*.'

Having learned this much about the acceptation of 'the Pill' in English, I found myself wondering: what about other languages? And (more to the point) does the informal epithet a nation chooses tell us something about a country's attitude toward birth control?

I started with the languages of which I have some reading competence. French, Spanish and Italian were unexceptional variations on the Rolph-Huxley theme: 'la Pilule,' 'la Píldora' or 'la Pillola.' But not so in German. Under 'Pille, die,' my **Duden** contains the pithy '*Arzneimittel in Form eines Kügelchens*' [medicine in form of a small ball]. For the more pregnant personal meaning of 'the Pill' or its international equivalents, I had to turn to the letter 'A' in my **Duden** to find '*Antibabypille*' with its laborious definition '*empfängnisverhütendes Mittel in Pillenform auf hormonaler Grundlage*' [contraceptive agent of hormonal basis in pill form—no charming *Kügelchen* diminutive here!]. Why the bitter, almost brutal, '*Antibabypille*'? Did the Church preempt this linguistic terrain before the journalists could settle on the pithy '*die Pille*'?

Steroid oral contraceptives were never designed as agents against babies. Since the Pill acts on the body of a woman who is not pregnant, there is no baby involved at all: the ultimate 'pro-woman' development, perhaps, but unless you imagine a woman's interests as somehow 'antibaby' it is hard to understand how such a term could come into use. Personally, I have no hesitation in calling it the 'pro-baby Pill,' because its ultimate purpose is to assure that every child is a wanted child. With the German propensity for complicated words, 'the anti-unwanted-child-Pill' would not be an unprecedented coinage, but I doubt if even the **Duden** would accept as an entry 'die Antiunerwünschtebabypille.' While most public media in Germany and Austria still seem to be chained to the 'Antibabypille' convention, most street German these days has fortunately dropped the pejorative 'anti-Baby' prefix. Is this just another manifestation of the linguistic Coca-Colanization of the present German generation or is it a token of a more realistic attitude in a country where the Pill—anti or not—has become the most popular method of birth control?

There is another reason for striking '*Antibabypille*' from the German vocabulary. It supports the common misconception—by no means limited to Germany or Austria—that the Pill is responsible for the rapid drop of birth rates in the industrialized nations, now that all of Europe but Albania and Malta has fallen far below simple replacement family sizes (2.1 children/family). This simplistic view must be challenged. Japan—a country with a disturbingly low birth rate—legalized the Pill only in 1999. Italy—the country with one of the lowest birth rates in Europe—reports one of the highest uses of coitus interruptus. Clearly, it is not a given birth control method, but the parental motivation for limiting the size of the family, that is the controlling factor. Historians undoubtedly would point out that this is a pre-Pill phenomenon: during much of 1870–1945, birth rates in industrial Europe were well below replacement level. It was only the post-war years that saw a worldwide reversal among industrialized nations.

Having once started on the slippery slope of looking to language for a key to societal attitudes toward birth control, I slalomed through languages as diverse as Swahili, Russian, Indonesian, Arabic, Chinese (Mandarin) and Hindi, but found only that the local phrase invariably meant 'anti-pregnancy tablet' or a minor variant thereof. I would like to interpret this as meaning that, for three fourths of the world's population, oral contraceptives are relatively unburdened with moral linguistic baggage.

II

Would that the rest of us were so lucky! The four decades since 'the Pill' became a household word have witnessed a veritable diarrhea of books and articles on the societal effects of the Pill, ranging from the unreservedly laudatory to the totally condemnatory, with every possible nuance in between. I myself have opined all too frequently on the subject, and still do in a formal course, 'Gender-specific Perspectives on Birth Control,' that I have taught since the early 1970s in various guises at Stanford University under the auspices of the Feminist Studies and the Human Biology Programs. The Pill has now been around long enough that defensive or proprietary comments no longer have much effect on its use. The Pill has become part of our social fabric; I shall leave it to others, mostly modern women, to judge its role within the perspective of their own lives. In my current reflective mood, rather than adding another bucket to that inexhaustible fountain of contention, let me instead summarize some of the critical commentary, starting with the feminist ideological opposition of the 1960s and 70s (now largely dissipated) and ending with the still-prevailing condemnation on the part of various shades of fundamentalists. As a social history of the past several decades, such a catalog leaves surprisingly few subjects uncovered.

In the 1960s, for obvious and deserved reasons, initially having nothing to do with the Pill, women became increasingly articulate about their social and economic roles in contemporary culture. One of the chief emphases of the early influential books of the modern feminist movement was the urgent need for improved *female* contraception. Simone de Beauvoir's *The Second Sex* stated explicitly, and Betty Friedan's *The Feminine Mystique* implicitly, that a liberated woman must be in control of her own fertility. Decades earlier, Margaret Sanger had offered the following definition of Feminism (her capitalization) in her autobiography (1938): 'Women should first free themselves from biological slavery, which could best be accomplished through birth control.' Probably most women will agree that the Pill heavily contributed to the achievement of that aim, but at the same time the convenience of the Pill and its wide acceptance by women gave many men the excuse to abandon their own responsibility. (It took the pandemic of HIV/AIDS and other sexually transmitted diseases to reverse that trend by making condoms once again popular, and even there the motivation isn't reproductive responsibility so much as—in the disingenuous words of the old vending machines—'the prevention of disease only.') But during the early years of the Pill's introduction, an informed and highly motivated minority of women—primarily North American and, by world standards, exceedingly affluent—unleashed an indictment of male domination that, in its attempt to account for the full scope of that oppression, swept up the Pill in its general charge. In so doing, these early feminists, for whom the politics of gender were paramount, claimed to speak for women all over the world. As usual, the rich were oblivious of the real problems of the poor: the contraceptive counterpart to 'let them eat cake' became 'let them use a diaphragm,' in facile disregard of the fact that millions of poor women in the Third World lack even a place to store a diaphragm, in addition to facing cultural barriers to handling their genitalia that sophisticated American or European women and men have difficulty imagining.

It angered those eloquent women that men had conducted virtually all the original chemical, biological, and clinical work on the Pill. And rather than seeing this situation as yet another instance of the exclusion of women from many areas of science and medicine, they saw a sexual conspiracy focused on an intimate aspect of their own sexuality. When the first large-scale, post-marketing, epidemiological studies documented some of the Pill's less obvious deleterious side-effects, women who earlier had objected to their use as human guinea pigs asked, 'Why wasn't the Pill tested more thoroughly?' The issue is a complicated one, but its salient feature is really not so much a matter of fact as of perception: given that most experimenters at that time were men, the development of a female contraceptive could only exacerbate women's feelings of being helpless and exploited.

There is another issue that makes this question even harder to resolve, which is the almost universal failure of our society to understand that terms like 'safety' or 'risk' can never be absolutes. There is no state of perfect safety; all decisions, whether they are to take a pill or not take a pill, or to use a car or an airplane, are choices between *levels* of risk. But try telling a white-knuckle flyer he's safer six miles up than at the wheel of his car; or a woman hearing tales of breast cancer and blood clots that she's safer on the Pill than pregnant! 'Safe,' in medicine, never means anything more than 'safe under the conditions of labeling.' Least appreciated is the fact that the safety of any vaccine or any drug to which a person will be exposed for long periods of time can only be determined by large-scale, post-marketing surveys, 'experiments,' in fact, in which the test population includes everyone who uses the drug. To proceed any other way would be to delay the introductions of new medicines by decades—and, if the drug is any good at all, to rack up a terrible toll in lives lost while treatments are delayed. Only medicines used in the short term to treat acute conditions can be effectively screened for most side-effects during the pre-marketing, clinical test phase. All of which sheds some light on why it took so long to lower the initial high dosages of the progestational and estrogenic ingredients of the Pill: it must be remembered that abortion was completely illegal at that time; experimenting with lower dosages might well have led to higher failure rates for which no alternative could be offered to the women on whom the new dosages were tested. Under those conditions, the then unknown risk of side-effects seemed preferable to the definitely undesirable risks of failure.

So when segments of the press, who had first promulgated the virtues of the Pill with naïve exuberance, began to sensationalize every new side-effect with headlines like 'Pill Kills!,' at a time when congressional committees blamed the FDA for every real and imagined oversight, no wonder women considered themselves the victims of a sinister cabal of avaricious drug companies and incompetent bureaucrats, whose technological output was peddled by members of what was then the most patriarchal profession of them all: gynecology. This conflict of cultural, technological and political forces found expression in a variety of forums, typified in the December 1970 issue of *Science for the People:*

> How is birth control practiced in our society? . . . We go to a doctor and lowering our eyes, embarrassed at our dependency, with a mixture of fear and anger we stumble through that horrible sentence, 'What do I do not to get pregnant?'

> Remember, we are asking this of a male doctor, behind whom stands the whole power-penis-potency complex (PPP). What do you think he's going to tell us? Right! 'Get high on our latest special, the PPP's Pill.' Great new wonder drug! It launches frontal attack on the pituitary gland and 'saves us from pregnancy' in exchange for a two-page long list of side-effects . . . which our male pharmacist or male doctor threw in the waste basket, and which we will never see. What we do see are little booklets from the drug companies decorated with roses, tulips, and peach blossoms full of reassuring babbling.

Early in 1970, such outcries and the drug industry's poor press came together in hearings before the US Senate Subcommittee on Monopoly of the Select Committee on Small Business, mercifully abbreviated as the 'Nelson hearings' after the subcommittee's chairman, Senator Gaylord Nelson, a liberal Democrat who had up to then done much to improve family planning in America and abroad. Senator Gaylord's public reputation at that time depended on what part of the public you belonged to. To the American pharmaceutical industry, he was the second coming of Torquemada; to the voting public at large he appeared more in the guise of Robin Hood, holding up the rich and powerful to uncomfortable scrutiny, if not outright loss of profits. In spite of the circus and klieg-light atmosphere, however, the Nelson hearings illuminated (for those seeking light on the subject) many little-understood aspects of Pill use and distribution, including many topics related to contraception. Hardly anyone at that time, however, predicted that this senatorial inquisition—now barely recalled—would become the pivotal event to push contraceptive research permanently into the minor leagues of biomedical research. The unexpected consequence was the extension of the Pill's life decades beyond the norm by restricting the arrival of new and better alternatives.

The Nelson hearings accomplished some good, notably in pressuring the FDA to demand that written information, with emphasis on potential side-effects, be inserted into every oral contraceptive package. Unfortunately, this requirement was fought by the medical profession and by pharmacists, who considered such inserts an usurpation of their professional function. As a result, once the industry's lawyers started working on the text of the insert, it turned into a densely printed, three-page document, which practically required a college degree in both legalese and biology to be fully comprehensible. I had always been in favor of package inserts for *all* drugs, from over-the-counter items like aspirin to prescription drugs. Nevertheless, I seriously question whether our present manner of listing every conceivable negative side-effect, however rare, does any more to convey important information to consumers than it does to protect the manufacturer against liability suits. Such overkill inserts can be frightening to the non-specialist, if not overwhelming. For instance, would not one view with caution, if not outright alarm, a pill reported as causing side-effects ranging from asthma, allergies, hives, edema, nausea, vomiting, disturbances of hearing and vision, anemia, mental confusion, sweating, thirst, diarrhea, and gastrointestinal bleeding all the way to occasional deaths? Yet these are the reported side-effects of aspirin. *Caveat emptor*, indeed!

In the end, the combination of anti-Pill women activists and Senator Nelson inadvertently caused the startling deterioration in contraceptive development, which began around 1970. It is difficult to avoid the conclusion that the bitter criticism aired and affirmed on Capitol Hill and the national press had something to do with the avalanche of liability suits that descended on the pharmaceutical industry; nor was the situation improved by the incessant rebukes handed out by the most underfunded regulatory agency of them all, the FDA, which had the none-too-salutary effect of terrifying it. The FDA's predictable response was a hypercaution with long-lasting consequences, the most important of which was the requirement, introduced in 1969, that steroid contraceptive drugs undergo long-term toxicology studies—unprecedented in any other drug category —for seven years in beagles and ten years in monkeys. These requirements were not modified until 20 years later as a result of overwhelming evidence presented by foreign regulatory agencies and the World Health Organization about the futility of such overkill mandates in animals such as beagle bitches, whose reproductive biology (*semiannual* canine heat cycles compared with human *monthly* menstrual cycles) and extreme sensitivity to female steroid sex hormones bears little resemblance to humans.

III

The withdrawal of most large American pharmaceutical companies (and subsequently most European concerns) from contraception research and development was accelerated by the barrage of liability suits over a period of about 15 years starting in the 1960s. Even though few such cases that went to trial in the US courts have been won by the plaintiffs, the cost of defending against such claims had by 1970 escalated to such an extent that it is often cheaper to settle out of court than it would be to take the case to trial, especially because of liberalized discovery rules permitting plaintiff's attorneys to demand tens of thousands of documents during the legal 'discovery' phase. Since the early 1980s, with increased epidemiological evidence of the health benefits of oral contraceptives (such as protective effects against ovarian and endometrial cancers and functional ovarian cysts) as well as reduced side-effects due to greatly decreased dosages, such suits have virtually dried up. But the lesson was never forgotten. The keepers of corporate accounts require little imagination to envisage an even more expensive horror scenario around any prospective male pill—one that a man might take for 30 or 40 years and then blame for his enlarged prostate or his waning libido. No wonder that, with the exception of one European firm, research on male contraceptives is presently shunned by all large pharmaceutical companies, whose affluence acts on litigation lawyers like a potent pheromone.

Academics and public health officials decry the withdrawal of the pharmaceutical industry from the contraceptive field and consider it important, as stated in an Institute of Medicine report in 1996, 'to show drug companies the massive need and potential market for new contraceptives.' That massive need may well exist, but not the potential market. Of the eight largest pharmaceutical companies in the world, only one still conducts some contraceptive R&D, and that of the 'me-too' variety; and only one of the top eight sells contraceptive drugs or devices. The pharmaceutical market, which has changed dramatically during the past decade, has spoken. It now focuses on blockbuster drugs dealing with diseases of aging or deterioration in the increasingly geriatric populations of affluent Japan, North America, and Europe. The needs of the poor pediatric societies of Latin America, Asia, and Africa cannot be met by market forces alone, which have a very weak track record where poverty is concerned. The situation is not improved by the dire competition for medical resources these societies now face. The four biggest killers worldwide remain what they were when the twentieth century dawned: acute respiratory infections, diarrheal diseases, tuberculosis, and malaria. All but one of these is curable by relatively inexpensive interventions, but in every case the diseases flourish for lack of effective vaccination, or difficulties in affording or delivering therapies where they are needed. Again consider the minute fraction of the huge R&D budgets of the top eight pharmaceutical companies dedicated to these fields. The moral: unmet burning societal needs do not necessarily equal financial returns. When it comes to the market delivering contraceptive innovations, the critics ignore the key point that the features of a truly novel contraceptive (say a contraceptive vaccine or a once-a-month menses-inducer pill) associated with major societal advantages (e.g. low cost and long duration for a vaccine; short action and minimal pill consumption involving 13 pills/year for a menses-inducer versus 250 or more pills per annum for current oral contraceptives), are precisely what would keep companies, which search for billion-dollar drugs used daily, from re-entering the contraceptive field.

IV

But let us return to the demands of those feminists, who wished some 30 or more years ago that a Pill for women had never been invented. What, if instead of having synthesized an orally effective progestin that ultimately led to the female Pill, we had made on 15 October 1951 another steroid that had become the basis of a male Pill? What if, instead of a daughter, we had contributed to the birth of a contraceptive son at a time when the threat of litigation did not yet exert such a dampening effect on contraceptive research?

In the 1960s, some feminist critics condemned the Pill as the ultimate invasion of a woman's body. Why, they asked, didn't these male scientists focus on their own genitals and leave our ovaries alone? I would maintain, in spite of the obvious imperfection of the Pill, that if that had happened, women would now be much worse off. Women's fertility would still have been largely dependent on the whim and pleasure of men. If I were a woman, I would rather tolerate the low incidence of potentially serious side-effects—the same kind of trade-off we make in every aspect of every daily life—than to give up one of the truly empowering agents of female emancipation. As a woman in the sexually loose 1960s, I would have been grossly unhappy, wondering whether the otherwise charming man who was about to share my bed was on his Pill (as he so fervently claimed), or whether my long-term partner had remembered taking it.

The attitudes expressed by feminist activists of the 1960s and 1970s have changed, although one can still find occasional anachronisms. One example is the contraceptive techno-Luddite fundamentalism expressed (in, of all years, the Orwellian 1984!) by one of the early feminist writers, Germaine Greer, who in her newest incarnation indicates that she had no use for the Pill and even denigrated the diaphragm in favor of coitus interruptus, the cervical cap (possibly the least used of contraceptive methods), and condoms. Another is illustrated by the *Daily Mail* of London—a conservative middle-market tabloid—that as late as 1999 saw fit to dedicate four-fifths of its entire editorial page to an article entitled 'As a woman I believe that the Pill could prove to be one of the greatest disasters of this century—socially, morally and medically.' The author concludes:

> The nightmare legacy of the Pill has, I believe, less to do with liberation than with ill health, the exploitation of women, and the decline of the family. It has also vested a concentration of power and control with the sex educators and family planners who have little time for honesty and informed debate. Isn't it time women saw the writing on the wall?

Fortunately, the current position of most informed feminist spokeswomen toward contraception in general, and the Pill in particular, reflects prevailing realities. Like the vast majority of women they want for themselves and for their partners more choices (including the Pill), to suit the personal and professional lifestyles of women working outside the home—perhaps the most significant and irreversible change in the status of women in the affluent (i.e. geriatric) countries of the world. They also want fuller and more up-to-date information on each method.

Yet just as women have entered every aspect of potential new contraceptive development—from converting the previously patriarchal discipline of obstetrics and gynecology into an increasingly female sub-specialty; from being represented at the WHO and especially in the US in substantial numbers in decision-making bodies dealing with contraception, such as advisory committees of the NIH, the National Academy of Sciences and the FDA; and currently even heading the entire FDA—their choices are becoming more limited and in my opinion will remain so. To me, this is the ultimate bitter pill.

V

Criticism from the left was distressing to many of the people involved in the early development of the Pill, largely because it was so unanticipated. Less surprising, but in many ways more troubling, was the range of attacks from more conservative elements in society, who saw in the Pill a symbol, if not an agent, of what they perceived as a pervasive moral decline. Many a critic complains that the introduction of oral contraception facilitated or even caused the loosening of sexual mores, which until then were largely enforced through the fear of unwanted pregnancies or of shotgun marriages. Is it not ironic that monogamous sexual behavior needs negative reinforcement in most societies, rather than relying on positive moral teaching? In the pre-Pill age, it was the fear of pregnancy; in the post-AIDS period, it has become the fear of sexually transmitted diseases. Yet it is difficult to tell whether the sexual revolution would not have been spawned anyway, given the concurrent appearance of the recreational drug and alternative cultures with their hedonistic messages and synergistic feedbacks. What can surely be posited is that if the Pill had not existed in the 1960s, the incidence of illegal abortions would have increased explosively. Thus, it is sad to read that the issue of abortion has politicized the subject of contraception to such an extent that the self-proclaimed Pro-Life advocates—in the US frequently also the group favoring the death penalty and opposed to gun control—all too often associate the large number of legal abortions since the early 1970s (when abortion was legalized) with the introduction of the Pill ten years earlier. In their obsession to declare pregnancy termination again illegal, the majority of these abortion opponents pay no attention to the virtual eclipse of the days when hundreds of thousands of women had illegal abortions, many self-induced or performed under dangerous and life-threatening conditions. A study published in the journal *Demografia* in 1969 estimated that in the 1960s, when abortion was totally illegal in the US and Europe, the number of illegal abortions in heavily Catholic countries such as Austria, Belgium and West Germany exceeded those of live births.

Yet fundamentalists are not the only ones attributing a cause-and-effect relationship to the Pill and the incidence of *legal* abortions since the mid-1970s. As recently as 1999, the respected social anthropologist Lionel Tiger argued that because pressure for liberal abortion intensified worldwide in the mid-1960s (i.e. *after* introduction of the Pill), the Pill was responsible for the putative worldwide increase in abortions. Not only do Tiger and so many other sweeping generalists ignore the rampant abortion rates in all of the Eastern European socialist countries as well as Japan, where the Pill was either unavailable or illegal, but they discount the enormous shift from the huge numbers of illegal abortions (which were generally not quantified) to the lower number of legal abortions that were reported openly. Hence Tiger's statement that 'only *after* women could control their reproduction excellently did they need more and more safe abortions' is in my opinion hogwash. What is wrong with the simple assumption that the desire for safer abortions by decriminalizing the procedure was an inexorable and logical demand of the women's movement, which flowered exactly during that period?

Before we allow the possible illegalization of abortion to happen again—a demand made by many fundamentalists in the US and elsewhere—we ought to learn from the experience of other countries—Romania, for instance—that have tried to outlaw abortion after it had been legal for years.

In 1965, abortion was legal and virtually free in Romania, whereas contraception was essentially unavailable; abortion, therefore, became the main vehicle for fertility control. The Romanian government was concerned not so much about the number or morality of abortions as about its falling birth rate at a time of relatively rapid economic growth. Overnight, that authoritarian government instituted a very restrictive abortion law, with the result, unprecedented anywhere in the world, that Romania's birth rate skyrocketed in one year from 13 to 34 per 1000. How jubilantly the men in Bucharest—and I use the word *men* with intention—must have congratulated themselves on the spectacular success of their anti-abortion policy.

Admittedly, this was policy, not politics, but the results were an almost unmitigated disaster. By 1968, the birth rate underwent an equally dramatic drop—from 34 back to 19 per 1000—

because it took less than two years for the abortion system to move underground. What the men in Bucharest had not taken into consideration, but could easily have predicted, was the increase in deaths associated with illegal abortions. By the time a decade had passed, this death rate had increased by a factor of 10; but the birth rate has not changed since 1972. Only with the downfall of the Ceausescu regime in 1989 and the renewed legalization of abortion did the maternal death rate drop again. Unfortunately, little was done in Romania to improve modern contraception with the consequence that abortions, though safer, are still the predominant method of birth control.

Now, if the self-anointed Pro-Lifers succeed to illegalize abortion in the US, there will unquestionably be a similar rise in the maternal death rate. This increase would, however, occur among the poorest women and would otherwise not be as widespread as many people predict. The reason is that, having had access to safe abortions for nearly three decades, American women will hardly permit themselves to resort once again to bent coat hangers. Initially, well over a million women might look for illegal but relatively safe abortions—which organized crime would take the opportunity of providing for a substantial fee. All but the most naïve will concur that criminalizing abortion inevitably will draw more criminals to the scene. Surely the lesson of Prohibition days should not be lost, nor that of the present drug-abuse problem. Unless we address ourselves to the underlying causes of such social problems, declaring them illegal is almost pointless, however virtuous we may feel in doing so. Why not instead focus on a common objective: how to make abortions unnecessary?

Much as laws against alcohol and drug abuse, however, are designed to protect the users, no anti-abortion law has ever been introduced anywhere for the protection of women. On occasion, as in Romania and earlier in Japan, the rationale has been blatantly demographic. Most countries, including the United States, have used these laws to enforce a standard of sexual morality; even today, the common objection that removing a few-weeks-old embryo is 'killing a baby' frequently reflects puritanical attitudes toward sex that the objector, whether sincere or not, feels should be imposed on others. Why do the vast majority of anti-abortionists make exceptions in cases of incest or rape? The answer is that the anti-abortionists consider a raped woman innocent and hence eligible for an abortion, while a woman who indulges in voluntary intercourse is judged guilty and hence doomed to an unwanted pregnancy.

This is as good a place as any to address at least briefly the problems of another pill, which, though not capitalized, has become even more contentious than oral contraceptives: RU-486 or mifepristone. Although chemically closely related to the 19-nor steroid skeleton of all oral contraceptives, there were two major differences—one of them a chemical substitution at position 11 of the steroid nucleus—that required the preparation of such compounds by total rather than partial synthesis. This was accomplished by Georges Teutsch and collaborators at the French pharmaceutical firm Roussel-Uclaf, where biologists discovered that such substances were powerful *anti*-progestational agents, with RU-486 being one of the most potent ones. Subsequent studies by Etienne-Emile Baulieu, André Ulmann and other clinicians showed that administration of this steroid shortly after a missed menses would cause expulsion of the conceptus in the majority of cases. By following ingestion of RU-486 with a prostaglandin (a non-steroidal fatty acid stimulating uterine contraction), the success rate could be increased to well over 90 per cent thus providing a medical alternative to surgical abortion. First introduced in France, and subsequently in the UK, Sweden, China and other countries, this remarkably safe chemical procedure has been used for the past decade in hundreds of thousands of instances under non-invasive conditions of privacy that for many women represented a powerful incentive for tolerating the side-effects of abdominal pain, nausea and bleeding associated with the drug regimen.

In the US, anti-abortionists and anti-choice groups promptly branded RU-486 as the 'French death pill,' threatening a boycott movement that terrified Roussel-Uclaf and especially its subsequent owner, the German pharmaceutical giant Hoechst. Financially, the market for a chemical abortifacient is small compared with the blockbuster drugs sought by today's pharmaceutical companies. Hence Hoechst decided not to pursue FDA approval in the US, but eventually—bowing to pressure from NOW (National Organization for Women) and other women's groups from the opposite side of the political spectrum—offered all patent rights to the non-profit and presumably non-boycottable Population Council. After 10 years of struggle—some legal, some bureaucratic, but most of it operational in nature—the FDA in September 2000 approved the use of mifepristone in the US. Identification of a manufacturer for the raw material proved to be the biggest hurdle, since no candidate in the US wanted to be exposed to the potential violent steps of the opposition. The eventual supplier (after the Hungarian firm Gideon Richter bowed out) proved to be a Chinese pharmaceutical company, which already was manufacturing the drug for the local market under FDA sanctioned laboratory practices and had been synthesizing oral contraceptives for many years before. The outrage of the anti-abortionists was understandable because RU-486 promises to decentralize the provision of abortion to a woman's bedroom, which can neither be bombed nor picketed. Filling a prescription for a few pills after seeing a physician in the privacy of his office is very different from having to pass the gauntlet of protesters in front of easily identifiable abortion clinics. But it remains to be seen what proportion of the 1.3 million women seeking annually an abortion in the US will turn to this chemical alternative.

By comparison to the violence generated by objectors to abortion, opposition to the Pill has not involved physical clashes. But just because some of that opposition verges 'only' on the irrational is not enough of a reason to ignore that component of the anti-Pill argument. Irrational fundamentalism is always worth engaging in debate, no matter how sordid the process, simply because otherwise it is capable by default of making its beliefs sound plausible to far too many of the voting public. It is like the evolution versus creationism debate: none but a professed masochist enters it, but woe to us if we have no such masochists on our side!

I shall illustrate the extreme anti-Pill position with a recent (May 2000) commentary from the religious Right, who are entitled to their opinion, but hardly to the vitriolic and unrelenting manner in which they wish to force their view of morality on everyone else. Father Richard Welch, the president of

Human Life International, which describes itself as 'the world's largest pro-life, pro-faith, and pro-family apostolate,' had this to say in an 11 May 2000 news wire on the fortieth anniversary of the FDA's approval of oral contraceptives:

> They should have been draping themselves in black and painting poison labels on the Pill's containers in mourning for the terrible damage done to individual women and to society by this evil creation. The dangers are nearly too numerous to list. Simply read the back of the insert where you can find the catalogue of possible side-effects, including death . . . And all of that doesn't address the most serious issue of all—the damage that the Pill, as the flag bearer of the contraceptive mentality, has done to our culture and our families. We are celebrating forty years of promiscuity, increased abortion, increased child abuse, increased divorce. Our television programs have become unwatchable because of their unrelenting sexual content. Our young girls are trained to be sexual toys. Our families are devastated . . . Is all of this possible because of a little pill? . . . The smallest, most innocuous seeming step—a tiny little pill—can cause endless damage. We must be ever faithful, ever alert, ever discerning. That is the only road to salvation.

Father Welch's simplistic attempt to blame all of contemporary society's real or imagined ills on the Pill is not worth rebutting, because he is not really talking about 'the Pill.' Father Welch and his trusting proselytes are against all contraception—perhaps an acceptable recipe for his personal salvation after his celibate life has come to a natural end, but an assured prescription for global disaster if applied to or forced on the rest of the world.

VI

The 1960s saw the flowering of three of contemporary society's most beneficial movements: environmental protectionism, consumer advocacy, and the women's movement. All three, though otherwise diverse, had this in common: they furthered their aims through the uniquely virulent American litigation system. Contraceptives and vaccines are obvious targets for such litigation, because they are not curative drugs to be taken by people already ill and hence tolerant of some side-effects; they are administered to the 'healthy' (itself a rather nebulous definition) to prevent a condition that the person may never get; it is hard to imagine a situation more likely to play into the popular belief that the normal situation is one of zero risk. Partly as a result of such vulnerability, vaccine makers in the US have been the beneficiaries of a self-funded, no-fault insurance program, instituted in 1986 through the National Childhood Vaccine Injury Act. Such a program, extended to the field of contraception, would be the single most important incentive for the re-entry of the pharmaceutical industry into the field. Unfortunately, there are differences in perception between vaccines and contraceptives—witness the political rhetoric that draws a distinction between 'innocent victims' of AIDS, those who contracted HIV at birth or from medical procedures, and those others, presumably culpable, that have contracted the virus by sexual intercourse. Such unspoken attitudes, which view sex as an act in which negative outcomes (pregnancy, disease) are somehow the result of divine justice, operate against extending any special protection to contraceptives. But the example of

AIDS may be rhetorical overkill here, repeating as it does a gesture all too common in this debate, which tends to focus on immediately observable consequences of health decisions—a specific disease like AIDS, or a frightening side-effect of a drug. What gets lost to such a short focus is the far more important, but also much subtler cost of our failure to develop an adequate contraceptive pharmacopoeia. The social and personal costs of an undesired pregnancy and of an unwanted child aren't good copy; as media sensationalism goes, it's just not 'sexy.' While successful and beneficial in many regards, in the field of contraception, the litigious approach pursued by diverse movements in the 1960s had the unexpected outcome of virtually demolishing the realization of a universally acclaimed aim: increasing the options for diverse, novel methods of birth control open to both women and men.

All we can expect well into the first couple of decades of the twenty-first century are minor modifications of existing methods: different delivery systems (such as skin patches) for currently used steroids; possible improvements in sterilization techniques and barrier methods; a more realistic reconsideration of the IUD; and more reliable ways of determining the infertile interval of a woman's cycle. (The motivation for that research, based on at-home methods involving a drop of urine or saliva, ironically got its impetus from determining the *fertile* period in women suffering from impaired fertility.) While such modest developments will not affect our total dependence on nineteenth- and twentieth-century approaches to birth control, it does not mean that birth control options cannot be widened.

The story of the 'morning-after Pill' (also known as 'emergency contraception' or 'postcoital Pill') is such an example. Though touted by the press as something new in 2000, its origin dates to the 1960s when it was noted that administration of orally effective estrogens, such as ethynylestradiol or diethylstilbestrol (DES), to rhesus monkeys within a few days following coitus prevented implantation of the fertilized egg in the uterine wall. If, however, implantation of the embryo had occurred, estrogen administration did not affect continuing pregnancy. Based on these monkey experiments, John M. Morris and Gertrude Van Wagenen at Yale developed an emergency contraceptive approach, involving administration of high doses of these estrogens within 72 hours following intercourse. Soon thereafter, the FDA approved the method for use in rape, incest, or even after unprotected intercourse. Health services in various colleges and universities had offered that medication since the late 1960s, but within a decade it fell into disrepute, primarily because of the discovery of a separate problem associated with DES: the appearance of a rare vaginal adenocarcinoma in young women who had been exposed *in utero* to DES when their mothers were prescribed this drug in the 1950s in the mistaken belief that it would prevent miscarriages. Administration of DES as a 'morning-after Pill' thus became contraindicated because of the risk that a woman may ingest that estrogen not knowing she is pregnant, and thus expose an unborn female child to the risks just mentioned. But within a few years (1974), the Canadian Albert Yuzpe reported that the same anti-implantation effect was also produced by the administration of high doses of conventional 'combined' oral contraceptives (consisting of an oral progestational steroid such as norgestrel combined with ethynylestradiol) within 72 hours of unprotected intercourse.

None of the pharmaceutical companies then selling the Pill wanted to promote that post-coital use and no one applied to the FDA for official approval. Hence its use remained an 'off-label' approach that the few physicians familiar with Yuzpe's regimen prescribed until the watershed year of 1995, when the Rockefeller Foundation's Bellagio Center in Italy hosted a medical expert panel on emergency contraception. Within months of that conference, the WHO added the Yuzpe combined oral contraceptive prescription to its list of essential drugs. Two years later, the FDA declared that high doses of six brands of commercially available combined oral contraceptives could be used for emergency contraception. By now, some specially packaged versions of either 'combined' or 'progestin-only' postcoital pills are on the market and some American pharmacies have started to dispense these to customers without a prescription but with counseling on their use. Since the side-effects associated with such short-term pill consumption are relatively benign (nausea, dizziness, headache, breast tenderness); since such emergency pills will not disrupt an established pregnancy; and since there is no evidence that such administration harms a developing fetus, a good case can be made that such 'morning-after' pills should be available without recourse to a doctor's prescription since their efficacy depends on taking them within 72 hours of unprotected intercourse. Suppose the coital act occurred on a Friday evening—a not uncommon weekend experience. Almost 60 hours will have passed before many of the most vulnerable women requiring such intervention could even get to a physician. That argument proved persuasive in the UK where in 2000 the government sanctioned such non-prescription use for women above the age of 16. In the USA, this enlightened approach has been implemented so far only in the State of Washington.

How widely such emergency contraception will be employed depends largely on the extent and nature of intelligent dissemination of proper information, preferably as part of sex education. Obviously, this method is not for everyone and especially not for women believing that life starts the second a sperm has penetrated the egg's membrane. Yet many people will share my belief that if used widely, emergency contraception is bound to reduce the incidence of abortions.

Why did I spend so much space on the morning-after Pill? Because from a chemical standpoint, it was created decades ago and required no mammoth biological or clinical efforts to bring it to its present state of perfection. What was missing until recently was an entrepreneur of Gregory Pincus's stature. If he were still alive, it would have been interesting to see how he would have addressed the area of emergency contraception.

Perhaps the only truly revolutionary technology on the near horizon will be another method—in-vitro fertilization—developed originally for the treatment of infertility that increasingly will also be employed by the fertile. Here again, we are not dealing with much basic science, but rather with applied research. The long-term storage of eggs and sperm, coupled with in-vitro fertilization at the time a couple wishes to conceive, could make early sterilization rather than contraception the norm among the affluent, especially when pre-implantation genetic screening of embryos becomes routine. Again, a bitter conclusion for many, but not for those millions of women and men who have chosen sterilization as a means of birth control.

The ironic conversion of the technologies of fertility to methods of contraception points to a larger force that we need to come to terms with before contraception can bring about the benefits it promises; what makes these conversions happen isn't so much technological change, but changes in the needs of the societies using those techniques, what I call the 'software' driving contraceptive 'hardware.' I have already argued that Japan, Western Europe, and North America do not depend on new methods of contraception for the solution of demographic problems. But what about the developing world: portions of Latin America, and especially Africa and the non-Chinese populations of Asia? Of the 11 largest countries in the world (with approximately 65 per cent of the world's total population), seven will double their populations in less than 45 years (with Pakistan and Nigeria doubling in less than 25)! It would be naïve to assume that advances in contraceptive methods (upgrading the contraceptive 'hardware') or even just improved access to existing methods of birth control would be the solution in the absence of concurrent major changes on the 'software' front—economic, educational and social improvements, notably in the status of women. Almost invariably, a change in women's education and cultural empowerment has carried with it positive economic consequences, which as history shows us lead to lower birth rates. Yet instead of seeing the desire for increased empowerment of women and the amelioration of population pressures in many non-industrialized 'pediatric' countries as synergistic aims, it is disconcerting to still encounter at the turn of the twenty-first century widely publicized statements reflecting the oversimplified perception that contraceptive development and promotion are primarily driven by the bureaucratic need to control population rather than a universal and individual wish to improve the quality of life.

As late as 1992, the Boston Women's Health Book Collective—an association focused on women's rights—stated categorically that 'We do not believe that overpopulation is a primary source of the world's problems.' For a person like me, who has seen the world's population triple within his current lifetime and may yet live to see it quadruple, it is the height of hubris for highly educated women in one of the richest countries of the world to make such an unqualified claim, pretending to speak, for instance, for women in the poorest countries, such as Pakistan, where the current population (150 million) will double in 25 years, thus packing a population that may approach in numbers that of the US into a country one-twelfth the size, with a commensurately far less efficient agriculture, infrastructure, and distribution system. Or to raise another question: how would the affluent women of Boston improve the quality of life in the poorest continent, Africa, whose current population of 800 million is estimated to grow to nearly 1.3 billion by the year 2025, while per capita food production has dropped consistently since the 1960s? Let them eat Maine lobster? In ignoring such questions, are we tacitly counting on the AIDS explosion in Africa to solve the equally scary problems of demographics?

VII

Starting in the 1960s, I was often asked, 'How do you feel about the social outcome of this work?' Depending on the circumstances, I may have grinned affectionately, shrugged my shoulders modestly, or even answered seriously that if I had to do things over again, there is little I, *as a chemist*, could or would have changed. Perhaps more relevant is the feeling that had I been a woman scientist, I would have been unashamedly proud to have been associated with the creation of oral contraception. I have no regrets that the Pill has contributed to the sexual revolution of our time and possibly expedited it, because most of those changes in sexual mores would have happened anyway—no matter what the *Daily Mail* or Father Welch or even the earlier mentioned Lionel Tiger may think.

But there is one question, frequently just implied, transmitted through look or inflection that turns me irritably defensive. It deals with the questioner's perception of the mint of money that supposedly ended up in my pocket as a result of my name appearing first on the list of the three inventors (the other two being Miramontes and Rosenkranz) of US Patent No. 2,744,122. I can give two answers.

One is short. In 1951, as a full-time employee of Syntex, my employment agreement contained the standard clause that every chemist working for a pharmaceutical company affirms to this day: for $1.00 and/or 'other valuable considerations,' the inventor agrees to sign all patent applications and to assign to the company all rights to any issued patent. 'Other valuable considerations' refers to the security of one's employment, the salary one receives, and possibly even a bonus or stock option, but never to a royalty based on a percentage of eventual sales. That would be reserved to outside inventors or other third parties.

I prefer, however, to answer the question by recounting my tussle with the *Berkeley Barb*, a muckraking tabloid that bit the dust in 1980. Three years prior to its demise, the paper published a long article criticizing the financial gains that had accrued to various university professors as a result of their association with the many biotechnology firms that had started to flourish in the San Francisco Bay area, and around Boston in the shadows of Harvard and MIT. Even though my own scientific research at that time had not impinged on the biotechnology revolution, the reporter quoted an apparently uncontaminated Berkeley professor to the effect that my academic position 'hadn't kept Stanford chemist Carl Djerassi from privately patenting birth control steroids he discovered under his own name for profit, even though he had discovered them while doing NIH-funded research. Perhaps significantly, Djerassi . . . used his own company to market such steroids.'

I had never before read the *Berkeley Barb*, but when several copies of this particular issue landed on my desk, I went ballistic. Their allegation, that I used government funds to feather my personal nest or that of my industrial employer, could (and, if true, should) have had a major impact on my academic career and on any further government funding of my academic research. I responded immediately, pointing to the public record: that the patent application on norethindrone was filed in November 1951, that the patent was assigned to my then-employer Syntex, that my Stanford University affiliation had started only in 1959, and that I had not filed a single patent application since that time. (While there is nothing illegal, or even improper, especially under current government regulations, in filing personal patents for inventions made with governmental subsidies in universities, I have chosen never to follow that practice.) I also added that I had never received any royalties for my work on oral contraceptives or for any of the other one-hundred-odd patents I hold for work performed while employed full-time by industry. Though the *Berkeley Barb* was not known as a paper likely to print apologies, in this instance they published a full-page retraction.

My reason for telling this story is that while the reporter printed an unequivocal *mea culpa* for not having checked the public record or interviewed me, he did insist that he had both quoted the Berkeley professor correctly and been given the impression 'that Dr Djerassi's alleged private patents on birth control drugs were common knowledge in the scientific community.' In that respect, I believe the reporter to have been dead right. There is little I can do about that perception, which is caused by a mixture of academic naïveté and wishful thinking, often also tainted by professional jealousy. Perhaps I should have told the *Berkeley Barb* that the continuing acceptance of the Pill by millions of women all over the world is the most valuable and also sweetest 'consideration.' And it would have been true. Mostly.

Chapter 4

The view from Tokyo

There is no ideal contraceptive. Differences in life-styles, sexual mores, the status of women, religious views, and many other factors guarantee that what is good or acceptable to one person is questionable or totally unacceptable to another. No wonder that use of the Pill and IUDs, the two most popular reversible contraceptive methods worldwide, differs dramatically from one country to another, and even among regions of a single country. None is more interesting in that regard than Japan.

Let us start, however, with Europe: the Pill is very popular in Holland and the United Kingdom, yet the former has one of the lowest and the latter one of the highest rates of teenage pregnancy. Is that a reflection of the high quality of sex education in Holland, or of the relatively lower overall use of contraceptives among British teenagers, or is it a combination of both? Until the mid-1980s, Italy had the lowest Pill use among Western European countries (this has now changed) and the greatest reliance on *coitus interruptus*, yet it had already approached one of the lowest birth rates in the world. Furthermore, in Western Europe, sterilization is lowest in Italy and highest in the UK and

Holland. Are Italian men more skilled at *coitus interruptus* than other Europeans?

While studies have shown that withdrawal before ejaculation is among the least effective methods of birth control, it nevertheless can have a significant demographic impact, especially when legal abortion is available in case of failure. The gradual reduction of the population growth-rate in Europe during the nineteenth century was to a large extent due to the widespread use of *coitus interruptus*. The founder of the British birth control movement, Francis Place, provided a particularly charming graphic prescription at the beginning of that century:

> The most convenient and easy, as well as the most effectual method is for the man at the moment of spending to throw himself on his *left* [emphasis added] side by which motion he not only in some measure extricates the part, but gives it also a slanting direction with respect to the woman, so that the seed being thrown not directly but in a side-long manner it is perfectly impossible for the womb to receive it.

Place's emphasis on the man's throwing himself on the left side implied that he was right-handed, thus using his stronger arm to push himself away from the woman. One wonders, tongue-in-cheek, if the technique was less effective for left-handed men trying to follow Place's advice to the letter. Or if Italians are more likely to be right-handed than other Europeans.

In the case of IUDs, national differences are even more strongly marked, and more easily explained. Although a highly effective and inexpensive method, the IUD has become a dirty word in contraception in the US because of the Dalkon Shield fiasco of the 1970s. In China, on the other hand, at least 45 million women depend on IUDs for birth control, compared with 7.6 million for the Pill. While many Chinese women still use an outdated version of metallic IUDs, European women have a range of plastic and medicated options that make IUDs an acceptable choice for many of them. Here the explanation is simpler. The virtual absence of IUDs in the US is a reflection of the devastating effect litigation can exert, while wide use of an outmoded IUD in China is a reflection of that country's authoritarian (though demographically effective) birth control policy.

Why does any of this matter? I prefer to answer that question through the metaphor of computer hardware and software (despite the unavoidable sexual innuendoes associated with these terms). Hardware, for me, are all the actual techniques used by people—steroid contraceptives, IUDs, abortion, condoms, sterilization, and related methods—in other words, the technology of contraception. Software covers the more difficult political, religious, legal, economic, and socio-cultural issues that individuals and, ultimately, governments must resolve before birth control hardware is employed—the ideology of contraception. Even the most sophisticated computer is of little use in the absence of effective software, and we all know what happens when hardware and software don't agree. In the end, it is the motivation of individual couples that determines the success or failure of any contraceptive technique, and it is here that the compatibility of hardware and software—among other things—comes into play. Once a decision about ideal family size has been made, the choice for contraceptive hardware may well be based on accessibility and affordability, but only after numerous other software hurdles are overcome. Japan—a powerhouse in computer technology—is a marvelous case for examining the importance of contraceptive software. Why, for instance, was Japan the last industrialized country in the world to allow women access to the Pill? The reasons are entirely of a software nature and are worth examining.

I

For me, the story started in early 1958, when Dr Edward Tyler of the Los Angeles Planned Parenthood Center, Dr Alejandro Zaffaroni, my colleague at the Mexican pharmaceutical company Syntex, and I flew to Japan under the auspices of the Japanese Pharmaceutical Society to present lectures on Syntex's then-new drug norethindrone as an orally effective progestational agent for menstrual regulation and for contraception. Our purpose was to arrange for commercial distribution of norethindrone in Japan. In a remarkably short time, Shionogi & Company, a large pharmaceutical company located in Osaka, became Syntex's Japanese distributor, and soon thereafter received Japanese government permission to market norethindrone, specifically for the treatment of 'menstrual disorders' (in addition to its more familiar use as a contraceptive, the progesterone-like activity of norethindrone can help in the treatment of irregular, painful, or otherwise troublesome menses). Around the same time, G.D. Searle & Co, Syntex's American competitor, also arranged for its product, norethynodrel, to be marketed in Japan for the same non-contraceptive clinical indications.

The initial dosages of norethindrone and norethynodrel recommended for treatment of menstrual disorders were roughly ten times those subsequently found to be effective for preventing ovulation (i.e., for contraception). Similarly, the estrogenic component of these preparations—the ingredient added for menstrual regularity but also responsible for some serious side-effects—was much higher. In other words, while effective at their officially-approved use, these drugs were also more than adequate as contraceptives. These 'high'- and 'mid-dose' therapeutic pills have remained on the Japanese market for 40 years—a point that is crucial to my subsequent story.

Given the rapid government approval of norethindrone and norethynodrel for the one purpose, I was convinced that Japan would be one of the first countries after the US to approve these steroids as oral contraceptives. There were certainly no religious obstacles. Indeed, on the face of it, Japan should have been the ideal country for early introduction. Only 30 years later, when I delved into the sociopolitical aspects of birth control in Japan, did the naïveté of my earlier assumptions become clear. Particularly helpful were subsequent research projects with two Japanese graduate students at Stanford, Mariko Jitsukawa and Hiromi Maruyama, the results of which crystallized for me the ambivalence of Japanese women toward that subject. Subsequent publication of our findings caused some stir in Japan—but not enough to make any difference.

First, the historical antecedents: Before World War II, all methods of birth control were illegal in Japan—even condoms, which, however, were provided for hygienic purposes to soldiers sent overseas. With the loss of the war and the end of any possible territorial extensions, the Japanese government instituted a strong anti-natalist policy, without, however, providing any hardware solution. As a result, illegal abortions became rampant, which prompted the Diet to pass the Eugenic Protection Law in 1948 which codified legal justifications for abortion, without, however, abolishing chapter 29 of the 1907 Japanese Penal Code that defined abortion as a crime. To compound this ambiguity, 'economic hardship' was added in 1949 as another permitted reason for surgical termination of pregnancies. Over the course of the next ten years, Japan's remarkable drop in the number of births mirrored almost exactly the equally dramatic rise in quasi-legal abortions (over one million annually by 1954), virtually all of them provided by the private medical sector.

The four-year gap between the availability of legal abortion (1948) and the first approval of contraceptive marketing (condoms in 1952) was the key factor for converting Japan into the first industrialized nation after World War II to depend on abortion for officially sanctioned birth control. Only in the early 1960s did the Japanese begin to reverse this trend with the wider spread use of the two contraceptive methods that until 1999 represented the entire contraceptive armamentarium of Japan: condoms and the *Ogino* calendar rhythm method (named after a Japanese physician). One would have thought that such a paucity of options would create a fertile (so to speak) breeding ground for adoption of an alternative method—specifically the Pill—that would reduce recourse to multiple abortions by so many women.

Indeed, in the early 1960s, *Kosei-sho*, Japan's FDA, was on the verge of recommending approval of the Pill. Before taking any action, however, the agency was faced with the thalidomide tragedy (in which women who took this widely-prescribed sedative during early pregnancy gave birth to children with serious birth defects) and several other drug-toxicity cases unrelated to contraceptives. The effect was to make *Kosei-sho* extraordinarily cautious. Ever since, safety rather than efficacy has become the watchword for the Japanese authorities. No wonder Japan now boasts more questionably efficacious but otherwise harmless drugs than any other modern country (including a putative anticancer remedy from shiitake mushrooms, selling at the rate of over half a billion dollars per annum, which no other country has approved).

Two other issues—associated, ironically, with the unsurpassed efficacy of oral contraceptives—kept the Pill (as well as IUDs) from the Japanese market, although officially the delay was attributed to a concern over side-effects. One was the opposition of the segment of the Japanese medical profession that provided abortion services and feared the loss of substantial income if the Pill were to be used widely in Japan. The size of this income can be judged from the fact that most Japanese physicians charge for abortions outside the national health insurance system, and do so at rates that are considerably higher than American rates—ranging from US$1000 (first trimester) to US$2000 (less frequent second trimester abortions). Such annual income, charged outside the national health insurance system, easily exceeds half a billion dollars, generating revenues that are frequently not reported to the tax authorities. In addition, the government feared that approval of the Pill would also make it more readily available to the unmarried young, thus fostering premarital sex. Such a concern is consistent with Japanese mores: its youth supposedly exhibits one of the industrialized world's lowest rates of premarital sexual activity.

In spite of over 30 years of clinical experience in the United States and Western Europe and subsequently in most other parts of the world, *Kosei-sho* still refused to accept the relevance of these data to Japanese women. Finally, after completion of a perfectly redundant clinical study in 1987 of low-dose oral contraceptives with a group of 5000 Japanese women volunteers, *Kosei-sho* slowly moved to follow the advice of a specially appointed advisory group of 12 men who recommended approval. (The absence of female members prompted the following editorial comment in the *Asahi Evening News* (22 December 1986): 'All 12 members of the Pill research group of the Health and Welfare Ministry were men. We do not know why the team was dominated by males, but this strikes us as strange.')

By early 1992, just when the media and the pharmaceutical industry anticipated final action, approval was again withheld, this time on the grounds that it might lead to reduced condom use at a time when AIDS prevention had finally registered on the Health Ministry's slowly-moving radar. In one stroke, the Japanese contraceptive clock was set back to where it had stopped some 30 years earlier. Many observers attributed this action not so much to any newfound preoccupation with AIDS but to other, more familiar reasons: the already-mentioned concern about the potential loosening of sexual mores, the opposition by condom manufacturers and abortion providers fearing reduced income, and the widely publicized low birth rate. The last argument is especially specious, since it implies that introduction of the more efficacious oral contraceptives would reduce the birth rate even further. Approval of the Pill would hardly do that—that demographic boat has sailed years earlier—but it certainly would reduce the 800,000–1million annual abortions.

II

To present the Japanese situation as a simple case of male imposition of law on an unwilling female population would be to simplify, however. Japanese women are themselves ambivalent toward modern contraceptives. Japanese feminists approach the issue in a manner reminiscent of some American feminist opponents of the Pill in the 1960s and 1970s. They emphasize the side-effects of high-dosage pills (which the rest of the world long ago abandoned for the much safer lower-dosage oral contraceptives). In addition, they object that the Pill would force them into a daily medication regimen, a form of contraceptive overkill in a country with a rather low coital frequency among married couples. (A fact of Japanese life so familiar that Japanese media have even coined a term for it: '*Sei-ai: Sei no nai kankei*', meaning 'Quiet love: relationship without sex,' but also implying that factors other than sex can lead to meaningful marital relations). Many of the women also object, not unreasonably, that the Pill causes them to assume full responsibility for birth control, whereas condoms require men's cooperation.

However, these opponents ignore the current use by possibly one million Japanese women of the high- and mid-dosage progestin-estrogen combination that *Kosei-sho* approved decades ago. Officially, such a regimen is only to be applied therapeutically for the treatment of gynecological disorders, but as many as 80 per cent of these users actually employ them solely for contraception. This means that hundreds of thousands of Japanese women may be consuming a contraceptive Pill without having recourse to the lower dosages with their well-documented advantages. Even more important, since their use for contraception was not officially sanctioned, none of the Japanese steroid preparations available through 1999 contained any package inserts or other descriptive information warning a woman about the consequences of long-term consumption of such high-dosage Pills.

Yet most women in Japan did not object to this deplorable state of affairs persisting to the very end of the twentieth century, because oral contraceptives have a double image in that country. First, women are afraid to be regarded as *sukimono* (sexually hedonistic)—willing to risk their health for sex. While the lack of marital sexual life is openly discussed in a country where vestiges of rejection of female sexuality are still prevalent, *sukimono* is for many Japanese women a social stigma or a synonym for the typical female sexual object in male pornography with its long historical tradition in Japan.

Second, the vast majority of Japanese women associate oral contraceptives with negative side-effects, without, however, being able to describe any. Factual information about the Pill has been totally blocked because the Ministry of Education has been reluctant to promote sex education, which for all practical purposes is also a taboo subject among parents and children. This puritanical fear of responsible sex education in schools—compounded by the absence of sex education in the home—is only now slowly changing because of the desire to promote

AIDS awareness, but even under the new dispensation discussion of such matters hardly touches on the Pill. Even medical professionals were not supposed to distribute information on the Pill to the public since it was not officially approved. Therefore, the main source of information remains the media, primarily the abundance of women's magazines, where little, if anything, is written about modern, low-dose oral contraceptives —or injectables, or IUDs, or any other modern means of birth control.

These negative views of the Pill by Japanese women have persisted until very recently. According to the biannual Family Planning Survey of the *Mainichi Shimbun* newspaper, only 35 per cent of Japanese women favored approval of the low-dose oral contraceptive Pill in 1986—the year when the Japanese government started the first cautious steps toward legalization of the Pill—but even this timid favorable rating dropped to 22 per cent in 1992 when the media reported that approval was imminent. No wonder that the government felt no urgency. In the end, it was not until the very end of the century that *Kosei-sho* finally took the jump. Even then, the *Mainichi Shimbun* survey of 2000 indicated that 73 per cent of married (and 52 per cent of unmarried) women would not resort to the Pill for fear of side-effects.

And what caused that final decision in 1999? Was it some form of millennial epiphany? Ironically, the final push may well be attributable to Viagra, which was approved by *Kosei-sho within one year* of its introduction in the US for the treatment of erectile dysfunction in men. This was the same *Kosei-sho*, by the way, that had refused for over three decades to accept the relevance to Japanese women of foreign clinical safety data on low-dose oral contraceptives; when faced with the threat of erectile dysfunction, however, it had no problem with approving Viagra's use in record time, and without any studies in Japanese men! That, of course, may not be surprising in a country with a higher proportion of older persons than any other in the world, and a long-established double standard in sexual conduct.

It remains to be seen whether the long-overdue approval of the low-dose 'contraceptive pill' will reduce either the Japanese abortion rate or its dependence on the high- and mid-dose 'therapeutic' pills for contraception. True to form, *Kosei-sho* is not simplifying that transition: it has issued guidelines that require a number of clinical tests for consumers of the low-dose Pill that are not required for women taking the same drug in higher doses! The costs for these tests may exceed US$200—this in addition to the cost of frequent consultations required to satisfy the needlessly complex restrictions on prescription refills. And none of this is covered by health insurance. Even more remarkably, the new low-dose 'contraceptive' Pill is more expensive than the mid- or high-dose 'therapeutic' Pill. As a result of all these artificial barriers, younger women in particular seem reluctant to shift to the less risky variant. Interestingly, *Kosei-sho* did throw in a political sweetener by approving concurrently the use of a copper-containing IUD, thus doubling in one swoop the meager contraceptive choices available to Japanese couples. Even the long-delayed IUD approval is ironic, since one of the very first metal IUDs—the Ota ring—was developed back in the 1930s by a Japanese pioneer, T. Ota.

III

It is not enough to say that Japan's attitude toward contraception—and especially the Pill—simply reflects the sexual conservatism of East Asian society. Sex education in the home is not much better in China or Korea, yet Pill usage has been sanctioned for decades in these countries. Until recently, Japanese did not even have its own word for the Pill, just using the English 'pill' ('*piru*' within the Japanese phonetic system); Mandarin and Korean, on the other hand, have their own terms: '*Bee-Yun-Yao*' and '*Pi-Yim-Yak*,' both meaning 'pregnancy-avoiding drug,' whose Japanese equivalent is '*Hi-nin-yaku*.'

Each of these societies has its own cultural peculiarities, but Japan again serves as an example *par excellence*. Starting around 1960, the Japanese birth rate displayed a consistent annual rise until 1972, when the trend turned persistently downward. But in 1966—right during the middle of that major upward movement—the birth rate showed an extraordinary dip that persisted only for that year. An imaginary Martian observer of Japanese demographics might have concluded that a natural disaster of unprecedented magnitude—killing on the order of 500,000 people—had struck the country. Yet the explanation is less dramatic and more charming. The year of the 'fire horse' had arrived—a mythological event occurring every 60 years that carries with it the belief that daughters born in that year will bring bad luck, even death, to their future husbands. So why take a chance? But the following year, the Japanese birth rate displayed the greatest single jump since the beginning of the century! One may well wonder whether this scenario will be replayed in

2026—the next Year of the Fire Horse—by which time sex predetermination through *in vitro* procedures such as intracycloplasmic sperm injection (ICSI) may well have penetrated the existing Japanese barriers to assisted reproduction. Will there simply be no girls born in 2026?

Since sexual behavior and attitudes are such personal matters, and at the same time so much a part of each culture's identity, differences should be accommodated rather than fought. Only a contraceptive supermarket—a multiplicity of choices to fit the peculiar software aspects of a given society—will provide such accommodation. In the long run, such an accommodation will also raise the status of women. By making reproduction a matter of choice more than an expression of biological law, contraception inevitably generates in a society a change in how women are imagined: less as servants of that law, more as conscious beings qualified to make choices; less controlled, more in control. Add to that fundamental shift in consciousness such practical considerations as freedom to shape a career around childbirth, rather than vice versa, liberation from the considerable morbidity of repeated abortions, and a new official recognition of the obligation to attend to the full range of women's health needs, and the benefits to women of a wider range of contraceptive choices become obvious.

In that sense, approval of the Pill and IUDs are steps forward, although such opinion is certainly not universal. Within weeks of the news of Japan's impending legalization of the Pill, the *New York Times* saw fit to publish a long Op-Ed piece predicting dire consequences for Japan: a rise in illegitimacy, divorce, and single-parent families. The editorialist attributed all of these darker aspects of Western society since the 1960s to the Pill. The writer even deplored the 'sharp decrease in the rate of shotgun marriages in the United States between the early 1960s and the 1980s'—not implausibly credited to the higher contraceptive efficacy of the Pill—because 'at the beginning of this period, society enforced strict rules on young men who got their girlfriends pregnant.' The article continued:

> Japan (along with Korea) stands out among developed countries because it lacks many social dysfunctions that have plagued Western societies. In the decades after World War II, violent crime rates in Japan actually declined, during a period when they were rising rapidly in the United States and Europe. Divorce has increased only slightly in the past 40 years, while single-parent families remain very rare. There are no extremes of poverty, drug use or teenage pregnancy . . .

The author, having blamed all these social ills on the Pill, ignores the fact that the Pill has been legal for years in Korea—a country he lauds together with Japan. While blaming the Pill for all Western ills, would the author be prepared to credit the lower rates in crime, divorce, and single-families in Japan since World War II to the explosion of abortions in a country where, according to a survey conducted in 2000, close to 80 per cent of umarried women approve of abortion? Whom would the author blame for another item in that survey (conducted by the *Mainichi Shimbun* newspaper), which concerns the rise of sexual intercourse among Japanese teenagers, from 14 per cent to 25 per cent, during the 1990s (when the Pill was still illegal in Japan), whereas the reverse trend has been noted in the Pill-happy USA?

Not surprisingly, the author of that collection of *non-sequiturs*, Francis Fukuyama, is a man, who chooses to ignore that, since the advent of oral contraception in the 1960s, the status of Western women has also seen major change which has allowed them to enter male-dominated areas outside the bedroom, nursery, and kitchen. If that movement can also be credited to the Pill, I for one would be happy to accept such a compliment. Unfortunately, that also would be a gross oversimplification. In the end, it is a woman's right to demand equal opportunities in the work place and sympathetic responses by men willing to carry some of the home and parenting burden that will change social life in the twenty-first century. The Luddite view that the fear of pregnancy—so prevalent prior to 1960—is the only way to enforce marital or social stability should be countered with the assertion that the Pill *per se* is neither the cause of most social ills nor their cure. Human beings cause these problems. In other words, it's the software that counts!

Chapter 5

Sex and immortality

As part of the rite of passage from one millennium into the next, a large German newspaper invited a diverse group of persons, including me, to write on 'Immortality.' The accompanying enticement, 'write anything you want on that subject,' proved sufficiently tempting that I agreed, before realizing how infrequently I had really thought about that topic. I do not believe in life after death. Hence at the most primitive level, immortality is a non-subject in my book of life. But it did not take me long to realize that the human longing for immortality expresses itself in many different ways. For instance, I could easily make the case that among many research scientists, such desire is stimulated by the values of our tribal culture. Why else be so preoccupied with name recognition, with the urge for publication under one's own name? Why else the preoccupation to be first? I openly admit that I would love to be able to read my own obituary—to be the proverbial fly on the wall that can observe what others say about me. Indeed, I have carried that fantasy so far that I have published an entire novel, *Marx, deceased*, about this very wish.

I

Like most humans, however, I have also aspired to immortality at a much more fundamental level: by reproducing, and thus perpetuating my genetic make-up beyond my own generation—temporarily forgetting the basic biological fact that this 'genetic immortality' is only partial. Even ignoring spontaneous mutations, only half of one's DNA gets to make the ride; worse, the trip requires someone else's DNA for completion, and it is impossible to know which alleles will be subsumed in others. Yet, is that not why people continually refer to 'bloodlines' and to the perpetuation of the family names, and why most people wish to have their own biological offspring rather than adopting children? The attraction—the wish to see our physical self reborn, our exact genetic duplicate given a second, or third, or an endlessly repeating chance at life—is easy enough to understand: who has not wanted a second chance, a helpful *doppelgänger*, the ultimately reliable voting bloc? But the fear is perhaps more deeply rooted in our race: the name of *doppelgänger* conjures up the dread of the uncanny, of the endless repetition of the Sorcerer's Apprentice, even as it undermines the very definition of the individual, the basis of our identity. In fact, is that not the reason for the enormous attraction and enormous fear of successful cloning? Fear, that my *doppelgänger*, though bearing every one of my genes, goes off on his own way, punishing me for having burdened him with my egotistic genetic baggage.

With very few exceptions, the millions of different species on this earth, from insects and reptiles to fish, birds and mammals, copulate in order to procreate. What we create in that act is not primarily an extension of ourselves, however: it is an extension of the species merely. What strives for immortality is not the individual—that ambition remains beyond our grasp—but the genome. This general set of chromosomes, rather than the particular combination of brown hair and hazel eyes, musical talent and athletic ineptitude, is the ultimate beneficiary of all that sweat and struggle in the dark. It carries no personal stamp of knowledge. With few exceptions, most males other than humans do not actually know who their offspring are, nor do the fathers of most species have anything to do with the upbringing of the next generation.

Not so with man. Parenthood is driven largely by a deep, personal association with one's children, indeed by obsessive identification with them. It takes little imagination to relate the desire for parenthood to a desire for a form of immortality. Once we recognize this equation, many of the traditional attempts at regulating sexuality take on a new significance. Until recently, becoming a biological parent invariably meant to achieve successful fertilization of a woman's egg by a man's sperm through sexual intercourse. Many religions, Catholicism being a prime example, insist that sexual intercourse not only be monogamous, thus clearly defining the biological identity of the offspring, but also that it be sanctioned only if reproduction is its formal objective. Judaism, on the other hand, through its reference to the mother rather than father for purposes of identification, tacitly acknowledges the uncertainty of paternal credentials. But these attempts at confirming the identity of the offspring are not all that seems to govern our traditional sexual mores: it does not so neatly explain, for instance, the Catholic Church's disapproval of contraception, which seems at times reducible to the injunction, 'You cannot have sex just for fun.'

Yet denying the overriding influence of the pleasurable aspects of sex is illogical. The Church is not against 'natural family planning,' against sexual intercourse during the time of a woman's menstrual cycle when she is infertile because she already ovulated or has not yet started to ovulate. The injunction would then seem to be something subtler, along the lines of, 'You cannot have sex just for fun, unless there is some element of risking conception.' It was primarily the uncertainty of accurately predicting the infertile days in a woman's monthly cycle that earned such 'natural family planning' the sobriquet 'Vatican roulette.' But now, as new biochemical techniques enter the market that permit women to determine with near-perfect certainty whether they are in a fertile period of their menstrual cycle, such 'hormonally-based natural family planning' just becomes another form of deliberate birth control. Why has the church so far not prohibited it? Is it because its relatively high failure rate is not due to the inherent uncertainty of such hormonal tests, but rather to human frailty: the lack of sexual discipline of the couples in refraining from intravaginal intercourse during the 'unsafe' period of the menstrual cycle?

There are other interesting religious exceptions to the ostensibly exclusive reproductive function of sexual intercourse. Among Orthodox Jews, sexual relations are not permitted during menstruation or when even the slightest evidence of spotting is observed. But there are women, who show occasional spotting during mid-cycle and during ovulation—in other words women who are not permitted to copulate precisely during their fertile period because the odd drop of blood can still be observed on a white sheet. Such couples are not infertile, but clearly they will not become parents through ordinary intercourse. (This prohibition of intercourse during bleeding, of course, has less to do with subordinating sexual pleasure to reproduction than it has to do with rituals of taboo/abjection/misogyny.)

Or take the man whose sperm count is 1–3 million sperm rather than the usual 50–150 million sperm per ejaculate. A couple of million sperm sounds like a lot of sperm, but they are insufficient to effect normal fertilization. Such men are diagnosed as suffering from severe oligospermia and are functionally infertile. But it is now possible to fertilize a woman's egg with the sperm of her functionally infertile husband by various forms of artificial insemination, provided his sperm is first collected in a condom—a procedure totally forbidden to observant, Orthodox Jews. 'Thou shalt not spill your seed'—an elegantly worded prohibition against masturbation—is the source of the Orthodox Jewish disapproval of condoms. Yet the Jewish religion, like most others, favors procreation, and a modern chief rabbi in Israel found a compromise that seemed Solomonic in its wisdom: He punctured a condom with a pin, so that a small amount of semen could leak through the minute opening, thus claiming the theoretical possibility of fertilization while retaining 90 per cent or more of the semen for subsequent artificial insemination techniques.

But whatever the uncertainties and inconsistencies that may arise as a result of the uneasy relationship between ancient religion and contemporary science, the overall pattern remains clear enough: in emphasizing reproduction over sexuality, in assuring that the offspring actually does convey the parents' genetic material into another generation, religion is simply serving one of its central functions, the promise of immortality. But must this genetic function be related to sex? Some of the most startling developments in contemporary science, and the social changes that accompany them, have started to shatter the historically unquestioned connection between sex and reproduction. The ultimate consequences of such separation will be profound, and not the least of them will be our ability to control the very nature of our immortality.

II

To reach that exalted end, we must start on somewhat lower ground, and observe that man is the sexiest of all species on earth. Among the millions of species, only we have sex for fun. Only we—and perhaps a couple of others such as the Pygmy Chimp (Bonobo)—are able and willing to have sex 365 days of the year. In all other species, copulation is seasonally controlled, and directly related to the optimal time for fertilization and the rearing of offspring. According to some reproductive biologists, such as Roger V. Short, the fact that man is the sexiest animal on earth is responsible for the extraordinary size (in relation to body size) of a man's erect penis. Compare it to that, say, of a gorilla, which is at best the size of a human thumb. Why should we need such an absurdly thick, swollen object to deliver sperm into a woman's vagina—ostensibly the only biologically significant, *reproductive* function of a penis? Clearly, we do not. A very thin, pipette-like structure would do equally well, if not better. Roger Short argues that the thick, massive penis produces more pleasure in the female partner, who is likely to prefer such men—better equipped for pleasure—as mates. Evolutionary selection, therefore, favored men with larger, thicker penises. If that argument is valid, one might reach the conclusion that sexual pleasure in the female becomes one of the determinants of selection, and that pleasure in addition to fertility comes to determine female receptivity, which then determines the frequency/timing of human sexual behavior.

But you needn't take my word for it, not when there are so many numbers that make the case more powerfully than words ever could. According to the World Health Organization (WHO), *every 24 hours* there occur over 100 million acts of sexual intercourse resulting in approximately one million conceptions, of which 50 per cent are unplanned and 25 per cent unwanted. It is this last estimate—250,000 *unwanted* conceptions a day—that is responsible for the occurrence every 24 hours of approximately 150,000 abortions, of which 50,000 are illegal, leading to the deaths each day of 500 women. What these figures do not say is how much effort has gone into avoiding conception before the fact, nor do they tell the tale of unwanted intercourse, or intercourse occurring under the influence of alcohol or other drugs. But even without those numbers, it is clear that something in addition to reproduction is driving all this. If a quarter of the conceptions that do occur are unwanted (unwanted even in the face of an ideology that so clearly favors reproduction), so unwanted that women risk legal sanction, even death, to end them fully 60 per cent of the time, then clearly a significant percentage of these

100 million daily coital acts has little to do with reproduction or any desire to perpetuate the species.

The *possibility* of indulging in sex without reproductive consequences through the widespread use of deliberate birth control is less than 100 years old (although history records plenty of recipes promoted to accomplish that aim). The true *realization* of 'sex for fun' occurred only about 40 years ago with the introduction of the Pill and of IUDs (intrauterine devices) that for the first time totally separated the coital act from contraception. Women who used these were temporarily sterile, and thus could indulge in sexual pleasure without the fear of an unintended pregnancy. Clearly, all of the millions and millions of couples indulging in such intercourse did so without any desire for reproductive immortality. In principle, for millions of couples, the decision to reproduce became a deliberate choice rather than a form of reproductive gambling.

To achieve a total separation of sex and fertilization requires two components. The first is effective contraception: the virtual guarantee of not creating new life during sexual intercourse. But this by itself does not permit a complete uncoupling of sex and fertility: to reproduce, one must still couple. Until recently, that is. The second component is the extreme counterpart of the first: to create new life *without* sexual intercourse. Its dawn coincided with the increasing use of artificial insemination, injecting millions of sperm into a woman's vagina rather than depending on *in situ* penile ejaculation. This low-tech method, using a syringe or even a turkey baster, underwent a spectacular escalation in technical sophistication in 1978 in England through the birth of Louise Joy Brown. Louise was conceived under a microscope, where her mother's egg was fertilized with her father's sperm; the fertilized egg was reintroduced into the mother's womb after two days, and, following an otherwise conventional pregnancy, a normal girl baby was born nine months later. This technique has since become widely known as *in vitro fertilization* (IVF)—an event that has now been replicated several hundred thousand times through the birth of that many IVF babies.

When Steptoe and Edwards developed IVF in 1977 they did not set out deliberately to make possible the separation of sex from fertilization. They, as well as other clinicians, were focused on the treatment of infertility. Infertility is itself an ethically charged topic. To put it bluntly and brutally: why should one treat infertility? From a *global* perspective, there are too many fertile parents, hence there are too many children, many of whom no one wants. The course of world history will not change if no case of infertility is ever treated, but it will change dramatically if excess human fertility is not curbed. From a *personal* perspective, however, the drive for successful parenthood is often overwhelming. Infertile couples are prepared to undergo enormous sacrifices, financially, psychologically as well as physically, to produce a live child under conditions where nature has made it impossible. The question may well be asked whether the realization of parenthood by biologically infertile couples carries some ethical imperative—for or against.

III

The enormous ethical dimensions of the problem become somewhat easier to see if we consider the question of male infertility. This issue was addressed in 1992, when a group of investigators (Palermo, Joris, Devroey, and van Steirteghem) in Belgium published a sensational paper announcing the birth of a normal baby boy fathered by a man with severe oligospermia (insufficient number of sperm). This child was made possible through the invention of an IVF technique called 'intracytoplasmic sperm injection' (ICSI), in which a single sperm under the microscope is injected directly into a human egg. Whereas in the original English IVF work, the egg was flooded with millions of sperm (as in ordinary sexual intercourse), with ICSI the artificial insemination was accomplished with one single sperm. The technology that makes such a fertilization possible also allows a radical revision of the definition of infertility: ICSI can be applied not only to men with low sperm counts, but also to men who have *no mature sperm whatsoever.*

Such men suffer from an inherited condition of total infertility, called 'congenital, bilateral absence of the vas deferens.' The vas deferens is the duct connecting the testis to the urethra, and is the organ where sperm is stored and then transported to the urethra and expelled during ejaculation. Without the vas, there is no sperm available for fertilizing an egg; clearly a man with such a condition can never become a father through sexual intercourse. But note that the barrier to fertility in such a case is not absolute. In fact, even immature sperm possess all the genetic material necessary to pass on a man's genetic heritage to posterity: it is the machinery of mobility, and the enzymes that penetrate the egg's cell wall, that are lacking, because they are acquired during maturation. With ICSI, however, the machinery of the laboratory can supply whatever the sperm lacks: one can aspirate immature sperm directly from the testis and inject its DNA into an egg under the microscope. Such fertilizations have been accomplished; numerous such men have now become successful fathers! Is this acceptable? Does such an infertile man have the right to *demand* that such reproductive technology be made available to him? And does it matter what motivates such a demand? Does it make a difference if we imagine we are settling the ultimate fate of a child—a concerned party, certainly, but one whose concerns cannot be said to exist except as a consequence of our decision? Or if we are simply satisfying a person's desire to achieve immortality? And how does it change that question if we consider what we are actually doing here: making the uninheritable (i.e. genetic infertility) heritable?

This turns out to be more than an ethical quibble. In one case out of four, men with 'congenital bilateral absence of the vas deferens' are also carriers of the gene for cystic fibrosis. With ICSI, one can envisage a scenario in which such men could pass on to their offspring both infertility and cystic fibrosis, raising the specter of successive generations requiring ICSI in order to perpetuate their genetic immortality—an immortality compromised by a disease that brings a slow, early death.

At the time of writing, the first ICSI baby is only 10 years old, but in that interval over 10,000 ICSI babies have been born. I have felt that the questions this technology raises merit wider debate than the traditional venues of a journal article or academic lecture allow. That is why I have incorporated these questions into a play, entitled *An Immaculate Misconception*. Here is an excerpt from a scene featuring a discussion between Dr Melanie Laidlaw, a reproductive biologist and (in the play) the inventor of ICSI, and her clinical colleague, Dr Felix Frankenthaler, whom she had invited into her laboratory. After she informed him that she is almost ready to perform the first ICSI injection into a human egg (without, however, volunteering that she will pick her own egg for such experimentation), they debate the possible implications of this work beyond simply treating male infertility:

Melanie

If your patients knew what I was up to in here . . . they'd be breaking down my door. Men with low sperm counts that can never become biological fathers in the usual way.

Felix

My patients just want to fertilize an egg. They won't care if it's under a microscope or in bed . . . as long as it's their own sperm.

Melanie

You're focusing on male infertility . . . that's your business. But do you realize what this will mean for women?

Felix

Of course! I treat male infertility to get women pregnant.

Melanie

Felix, you haven't changed. You're a first-class doctor . . . but I see further than you. [Pause.] ICSI could become an answer to overcoming the biological clock. And if that works, it will affect many more women than there are infertile men. [Grins]. *I'll even become famous.*

Felix

Sure . . . you'll be famous . . . world-famous . . . if that first ICSI fertilization is successful . . . and if a normal baby is born. But what's that got to do with [slightly sarcastic] 'the biological clock?'

Melanie

Felix, in your IVF practice, it's not uncommon to freeze embryos for months and years before implanting them into a woman. Now take frozen eggs.

Felix

I know all about frozen eggs . . . When you rethaw them, artificial insemination hardly ever works . . . Do you want to hear the reasons for those failures?

Melanie

Who cares? What I'm doing isn't ordinary artificial insemination . . . exposing the egg to lots of sperm and then letting them struggle on their own through the egg's natural barrier. [Pause]. *We inject right into the egg . . . Now, if ICSI works in humans . . . think of those women—right now, mostly professional ones—who postpone childbearing to their late thirties or even early forties. By then, the quality of their eggs . . . their own eggs . . . is not what it was when they were ten years younger. But with ICSI, such women could draw on a bank account of their frozen young eggs and have a much better chance of having a normal pregnancy later on in life. I'm not talking about surrogate eggs . . .*

Felix

Later in life? Past the menopause?

Melanie

You convert men in their fifties into successful donors . . .

Felix

Then why not women? Are you serious?

Melanie

I see no reason why women shouldn't have that option . . . at least under some circumstances.

Felix

Well—if that works . . . you won't just become famous . . . you'll be notorious.

Melanie

Think beyond that . . . to a wider vision of ICSI. I'm sure the day will come—maybe in another thirty years or even earlier—when sex and fertilization will be separate. Sex will be for love or lust . . .

Felix

And reproduction under the microscope?

Melanie

And why not?

Felix

Reducing men to providers of a single sperm?

Melanie

What's wrong with that ... emphasizing quality rather than quantity? I'm not talking of test tube babies or genetic manipulation. And I'm certainly not promoting ovarian promiscuity, trying different men's sperm for each egg.

Felix

'Ovarian promiscuity!' That's a new one. But then what?

Melanie

Each embryo will be screened genetically before the best one is transferred back into the woman's uterus. All we'll be doing is improving the odds over Nature's roll of the dice. Before you know it, the twenty-first century will be called 'The Century of Art.'

Felix

Not science? Or technology?

Melanie

The science of . . . A . . . R . . . T [Pause]: assisted reproductive technologies. Young men and women will open reproductive bank accounts full of frozen sperm and eggs. And when they want a baby, they'll go to the bank to check out what they need.

Felix

And once they have such a bank account . . . get sterilized?

Melanie

Exactly. If my prediction is on target, contraception will become superfluous.

Felix

[Ironic]

I see. And the pill will end up in a museum . . . of twentieth century ART?

Melanie

Of course it won't happen overnight . . . But A . . . R . . . T is pushing us that way . . . and I'm not saying it's all for the good. It will first happen among the most affluent people . . . and certainly not all over the world. At the outset, I suspect it will be right here . . . in the States . . . and especially in California.

Felix

The Laidlaw Brave New World. Before you know it, single women in that world may well be tempted to use ICSI to become the Amazons of the twenty-first century.

Melanie

Forget about the Amazons! Instead, think of women who haven't found the right partner . . . or had been stuck with a lousy guy . . . or women who just want a child before it's too late . . .in other words, Felix, think of women like me.

ICSI raises many other ethical and social problems beyond those mentioned in the Melanie/Felix dialog. For example, now that the effective separation of Y- and X- chromosome-bearing sperm has been perfected, ICSI will enable parents to choose the sex of their offspring with 100 per cent certainty. For a couple with three or four daughters, who keep on breeding in order to have a son, the ability to choose a child's sex may actually prove a benefit to society, but what if practiced widely in cultures (such as China or India) that greatly favor male children over females? Even more complicated are the issues raised by sex pre-determination motivated for medical reasons. How would you judge the morality of the following concrete intervention realized in the year 2000?

Hemophilia is carried forward through the mothers, meaning that while male hemophiliacs do not transmit the disease to their children, daughters born from such fathers become carriers of the disease to their own sons. In Spain, such a family decided to use only male embryos during *in vitro* fertilization to avoid the danger of hemophilia in their eventual grandchildren, thus jumping a generation to stamp out the disease in that family.

Moving beyond sex-predetermination, what about the capability of preserving the sperm of a recently deceased man (say 24–30 hours post-mortem) in order to produce (through ICSI) a live child months or even years later—a feat that has already been accomplished? Here we have immortality with a vengeance. But what of the product of such a technological *tour de force*? Using the frozen sperm and egg of deceased parents would generate instant orphans under the microscope. The prospect is grotesque—yet does it take much imagination or compassion to conceive of circumstances where a widow might use the sperm of a beloved deceased husband so that she can have their only child? Or what about our infatuation with pets? Affluent dog lovers have preserved the sperm of their favorite dog to ensure the closest possible surrogate after their pet's demise. One such posthumously fathered dog was described in the *New York Times* as being 'truly our little king, and he looks just like his father.' According to that article, even that 'little king's' sperm was soon to be frozen for continuing the canine pet succession.

These issues are intrinsically gray; the technology occupies an ambiguous position, enabling us to enact our best and worst impulses, and the answers cannot be provided by scientists or technologists. The ultimate judgment must be society's, which, in the case of sex and reproduction, really means the affected individual. Ultimately, that individual is the child, yet the decision must be made before its birth by the parents—or more often than we care to admit, by just one parent.

IV

It is the nature of such questions that they resist convenient solutions, not least because of their tendency to proliferate faster than we can solve them. Whereas reproduction has historically tended to exemplify the law of unintended consequences, the addition of technology has given that law added force. Consider: until very recently, the onset of menopause was welcomed by many women as the release from continuous pregnancies caused by unprotected and frequently unwanted intercourse. But the arrival of the Pill and other effective contraceptives, coupled with the greatly increased number of women entering demanding professions that cause them to delay childbirth until their late 30s or early 40s, now raises the concern that menopause may prevent them from becoming mothers altogether. Whereas reproductive technology's focus during the latter half of the twentieth century was contraception, the technological challenge of the new millennium may well be conception (or infection, if one focuses on sexually transmitted diseases). In the long run, if the cryopreservation of gametes followed by sterilization becomes a common practice, contraception may even become superfluous. Melanie and Felix in the above fictitious dialog were hardly the first to express such speculation.

In 1994, in the scientific journal *Nature*, the cryobiologist Stanley Leibo and I addressed the deplorable prognosis for a new male contraceptive in the next few decades, given the total lack of interest in that field by the large pharmaceutical companies without whose participation such a 'Pill for Men' could never be introduced. This led us to propose an alternative approach, not involving the drug industry, based on a few simple assumptions.

Millions of men—admittedly, most of them middle-aged fathers rather than young men—have resorted to sterilization (vasectomy) and continue to do so. The procedure is much simpler and less invasive than tubal ligation in women. (Sterilization among both sexes has become so prevalent, that in the US, it is now the most common method of birth control among married couples, even surpassing the Pill.) Artificial insemination is both simple and cheap. Furthermore, among fertile couples, it has almost the same success rate as ordinary sexual intercourse. But most important for our argument, fertile male sperm has already been preserved inexpensively for years at liquid nitrogen temperatures. Therefore, provided one first demonstrated that such storage is possible for several decades rather than just years, some young men might well consider early vasectomy, coupled with cryopreservation of their fertile sperm and subsequent artificial insemination, as a viable alternative to effective birth control. Shifting more of that responsibility to men, at least in monogamous, trusting relationships, appeared to Leibo and me a socially responsible suggestion. I shall spare the readers a record of the resulting outcry—both by media and in personal correspondence—but a lot has happened in the intervening few years to make it much more likely that such a prediction will become fact rather than dramatic license within a few decades.

Although many may consider some of the scenarios raised in *An Immaculate Misconception* as 'unnatural' or worse, every one of them has now been realized or is about to be implemented. Take the question of post-menopausal pregnancies. In progressively more geriatric societies (for example, in Japan or Western Europe), where 20 per cent of the population is already or will soon be over the age of 60, and older people are increasingly healthier than they used to be, a woman who becomes a mother at 45 could raise a child for a considerably longer time than could a 20-year-old at the beginning of this century. Of course, motherhood at an older age is physically, psychologically, and economically suitable only for certain women, but at least the choice is now available in wealthy countries. It must be emphasized that this increased emphasis on artificial fertilization techniques and even surrogate parentship is a characteristic of the affluent, 'geriatric' countries. Even within these countries, the cost of such reproductive technologies (frequently not covered by insurance) is such that only the more affluent citizens can afford them. Three-fourths of the world's population is represented by the 'pediatric' countries of Africa, Asia and much of Latin America, where over 40 per cent of the population may be below the age of 15 and where the control of fertility rather than the treatment of infertility will remain the catchword for decades to come.

I have deliberately refrained from considering the implications of human cloning—the closest technological approach to immortality where precise genetic replication is achieved. Technically, I am confident that this will be accomplished in the not-too-distant future. But is it desirable? Cloning one's own organs for therapeutic organ replacement will certainly be feasible and in my opinion also beneficial. Cloning an entire human being is a different proposition. If we are honest, many of us will admit that we are less worried with what *we*, personally, would do with such a technique rather than how *they* would misuse it. (During a survey conducted in the USA in November 2000, 90 percent of respondents favored the cloning of pets.) The definition of 'they' covers the gamut from governments or dictators to evil Strangeloves or even just egocentric monomaniacs. It hardly takes an Aldous Huxley or a sophisticated geneticist to highlight the penalties of genetic inbreeding if such human cloning were carried out on a macro scale.

To the extent that biological parenthood is a form of immortality—admittedly one subject to mutational and hence evolutionary adjustments—IVF tampers with that as well. In the excerpt from my play, I allude to pre-implantation embryonic genetic screening, again a procedure primarily available to the affluent in the affluent countries. But soon, the entire human genome will be elucidated: given the many technically feasible methods of rapid genetic screening, what will keep prospective IVF parents from screening their own embryos so as to transfer only the 'best' back into the mother? Who will define 'best'? Few people will argue that prospective parents may wish to discard embryos that show the markers for Down's or Huntington's Syndrome, or markers for genetically transmitted cancers—prospective parents, who follow the motto 'better be tested now than diagnosed later.' How many parents are willing to judge the Colorado couple, whose daughter suffered from Fanconi anemia, a genetic condition preventing the production of bone marrow—a certain prescription for early death. With their second child, by pre-implantation screening of the embryos for the Fanconi anemia marker, they were able to guarantee that their next offspring—a son, named Adam— would be free of that disease and could serve as a compatible tissue donor for his six-year older sister. In fact, transplant of umbilical cord stem cells from the 'made-to-order' Adam caused the older sister to start making platelets and white blood cells on her own.

But where will the line be drawn? Short stature? Left-handedness? Big ears? As we move in the direction of tailor-made progeny, the gulf between the haves and have-nots is widening enormously. Perhaps the longing for immortality should focus purely on one's intellectual or societally significant accomplishments, rather than biological replication. In the next century, let alone millennium, love or lust for one's partner will continue to be the principal incentive for sexual congress. The decision to produce a child involves a different kind of love, ideally tempered by responsibility and serious reflection. The recent advances in contraceptive and reproductive technologies now permit us to make that choice, but whether we shall display the necessary wisdom and restraint remains to be proven.

Chapter 6

From the Pill to the PC

I

On 8 May 1983, the love of my life, whom I—a recent divorcé with a roving eye—had met on Valentine's day of 1977, announced with a tender thunderbolt that she had fallen in love with another man. We were through, she said unequivocally, though much more elegantly than I give it here. We were through—the words were truer than either one of us imagined. What was ending wasn't, however, our love. It was the life I had known until then, soon to be supplanted by a new, utterly unexpected turn away from scientific research.

My response at the time, however, was somewhat less exalted—typically male, in fact. How could she fall in love with another man when she had me? And how come I had no inkling? Humorless, self-pitying, the urge for revenge—charged with testosterone and adrenaline—led me in two directions. One was obvious; the other surprised even me.

The obvious one was new female companionship; the surprise was an outpouring of poems—confessional, self-pitying, and even narcissistic. It was a cathartic experience for someone who until then had never written a single line of verse. Like a typical scientist, believing that anything worth doing is worth publishing, I looked for the poetic equivalent of scientific journals. In no time, I discovered a plethora of poetry magazines, most unknown to me, which, to my utter surprise, all had the temerity to turn down my sobbing lines. But as humorless self-pity turned into more tasteful sarcasm, one editor bit. The first two stanzas of *Lingonberries Are Not Sufficient* from a 1984 issue of the *Cumberland Poetry Review*, describe what had brought us together.

> When we met
> I spoke and you laughed.
> Just laughed.
>
> Come live with me
> And write.
> I'll feed and launder you.
> You laughed and came.

It takes another stanza from my last published poem to indicate what took us apart.

> But he,
> Master of chemical mutations
> Whose alchemy touched millions,
> Could not transform her,
> Nor transform himself.

II

The 'you' in the poem was Diane Wood Middlebrook. An English professor at Stanford University, during our first year together, she not only spun tales and poetry, but also completed a book on *Understanding Modern Poems.* That elegantly slim volume got me sufficiently hooked on Wallace Stevens—the subject of her first monograph and of some of her Stanford seminars—that I attempted to use him as the ultimate authority on a contentious point of linguistic idiosyncrasy.

I am told that I still have residues of an accent—colored by my mother tongue (German, by way of Vienna). I learned English at the American College in Sofia, Bulgaria from a mix of British and American instructors, and their voices still compete with the sea of Slavic accents of my fellow classmates. It is only when I listen to my voice on the radio or on a tape do I realize that I speak no language—not even my mother tongue—accentless. But until I met Diane Middlebrook, I suffered under the delusion that my English was (at least) impeccably grammatical. After all, it is the language in which I dream.

'I think we should make reservations already now for Christmas in St Maarten,' I announced during our first month of living together.

'What do you mean by "already now"?' she laughed. 'It's redundant. It's probably German.'

'German?' I hooted, until I realized that she had simply displayed her innate stylistic *Fingerspitzengefühl.* Of course she was right: '*schon jetzt*' is what every sensible German-speaking *mensch* would say: a 'now' feels naked and incomplete without the security of an 'already.' But does that make it tainted English?

The question rankled. I enlisted my loyal and literate secretary's help. I asked Guynn Perry, an inveterate reader, to look out for 'already nows' in her belletristic browsing. She did not disappoint me, although I have to admit that I used her evidence with the nonchalant *droit de seigneur* of a science professor appropriating his graduate student's findings under the disguise of the first person plural. 'Look what *we* have found: Leo Tolstoy,' I announced to my resident English Professor. 'I know I'm going to die. You ask him. I feel it *already now.*' *Anna Karenina.* In fact, Anna's own words.'

Diane didn't even laugh. 'A literal translation from the Russian,' she said dismissively.

We—meaning my secretary—did not give up. 'What about this?' I asked a few days later. '*Already now* I feel, as when at the age of twenty I was going to a ball in the evening, that day is a space of time without meaning . . .'

Diane perked up until I cited 'our' source: *Shadows on the Grass* by Isak Dinesen. Diane never sneers, but she came close that time, the way she said: 'Danish, not English.'

I refused to give up. Instead, I raised the ante: If I could produce an irreproachably English use of 'already now,' would she compose a poem containing the phrase? When she agreed (I never determined whether it was amusement or exhaustion or what other reaction that prompted her assent), I produced an impeccable source. A few days after Tolstoy and then Dinesen had bit the dust, I handed over a Xerox copy of a page from her own Wallace Stevens's *Harmonium.* The page held the poem *In the Carolinas.* I began:

> The lilacs wither in the Carolinas.
> Already the butterflies flutter above the cabins.
> Already now the new-born children interpret love
> In the voices of mothers.

I might have gotten away with it. According to Walsh's *Concordance to the Poetry of Wallace Stevens*, which I had consulted, the word 'already' appears in only six lines of the poet's huge oeuvre. All I had done was simply splice the word 'now' into *In the Carolinas* by lifting those precious three letters from another poem printed in the same font. But (and I fear this is typical of me), I was not content with a single literary crime. I also presented her with a doctored version of Stevens's shortest poem, *To the Roaring Wind*, which I had lengthened by two words.

What syllable are you seeking,
Vocalissimus,
In the distances of sleep?
Speak it already now.

It is chancy enough to present a Stevens scholar with a tampered line, but picking this particular one was enough to make a risk into a certainty. Weeks earlier, I had started to call Diane 'Vocalissima' (and this has remained my term of endearment for her ever since). She saw through the fake immediately, but as a true Vocalissima she paid off a wager she had never lost by presenting me with the following gift I have always cherished, because it is so typical of her: elegant, generous, yet still unyielding.

Sonnet: Auspices (29 July 1977)

Already, now, the moon, Diana's sign,
Asserts herself among your darkened trees,
Abandoning the branches one by one.
Gaining the sky she claims your sleeping form;

Dreaming, you do not feel her fluent gaze
Rinsing your face with silver to the rim.

Four months. Is this a sign, this storm of light
Shedding its silent thunder in my blood?
Pagan through love I scan the moving sky
And make of every sound an oracle.
The future seems a tantalizing gift
The moon might open to her priestess here—

The moon is silent. Wind is merely sound.
And yet a benediction fills the night;
My voice, aloud, draws meaning from this hour,
Finding four syllables. Already. Now.

(Diane Wood Middlebrook)

Why do I include this shaggy-dog story at such a point in my autopsychoanalytical musings? Is it to confess a bulldog streak? Or is it offered as evidence for some innate linguistic spark, an ember that lay waiting a suitably inspiring breath? I dare not answer that last question, because 'linguistic spark' is too approving a term to claim on my own. I shall let my readers judge.

III

My prolonged exposure to Wallace Stevens during the preceding six years even caused me to indulge in some Stevensiana of my own, through five poems that eventually appeared in the *Wallace Stevens Journal*—hardly a household publication venue for an American organic chemist. The first sample's title comes from a line by Stevens and shows that the wound caused by our year-long separation had not yet healed:

Pure Scientist, You Look With Nice Aplomb

I, the scientist,
Read these words
And snort
Without aplomb.

Today
Is any scientist
Pure?
Especially to poets?

Is aplomb
Ever nice
To the spectator
Hiding envy?

Poet—don't you know
There are as few scientists
With aplomb
As there are poets?

As the number of my published poems grew, I started to take myself seriously enough to seek a publisher willing to publish a chapbook of mine. At the same time, however, I was sensible enough to recognize that it was too late in life to turn into a real poet: I realized that the chief value of this lyrical outpouring was therapeutic. I had reached the point where my poems had lost their self-pitying bitterness; to my relief, they had even started to acquire a touch of humor. Already a decade earlier, my research interests had expanded to the insect field. Now, as the discarded lover, I turned my arthropodan browsing into a score for my poetic lyre. The following excerpt from a 1984 scientific article by Izuru Yamamoto became the long epigraph to a poem in my chapbook's *Amour D'Arthropode* series:

> Erectin, a sex pheromone of the female azuki bean weevil (*Callosobruchus chinensis*) induces the male insect to extrude his genital organ for copulation. The male will also copulate with a glass rod, a tubular piece of aluminum foil, or any other surrogate doused with the pheromone. This presents possibilities for population control with an aluminum foil tube being a simple and satisfactory dummy.

Three variations on a theme by Callosobruchus

1
Blindfold me, Izuru.
Douse me with Erectin.
Then tempt me
With aluminum or glass.

2
Will Pepsi empties,
Empties of Perrier
Be rated G, PG,
Or X?

3
Merrill Lynch?
Buy me ALCOA.
Get me CORNING.
I'm bullish.

Had I turned the corner by using poetry as the balm for an abandoned lover's wound? Alas, the answer was 'not yet.'

IV

Finding poetry too constraining a vehicle for my narcissistic wrath, I decided to write a novel of unrequited and discarded amour, a clever *roman á clef* focusing on a terrible lapse of judgment on the part of an elegant feminist, who had dropped her eminent scientist-lover for some unknown littérateur. My masterpiece's title, 'Middles', I borrowed from Nora Ephron's *Heartburn* (1983), a *cri de coeur* I heard with a sympathetic ear. Ephron's novel was a piercing literary stiletto serving as the *pièce de revanche* against her former husband Carl Bernstein, who had abandoned her for a newer model. It is true that Ephron wrote: 'I insist on happy endings; I would insist on happy beginnings, too, but that's not necessary because all beginnings are intrinsically happy . . . middles are a problem. Middles are perhaps the major problem of contemporary society.' But as my novel's *objet de revanche* was meant to be Diane Middlebrook, I was embarking on roman-á-cleffery at an extraordinarily transparent level by simply using my novel's title as an unsubtle reference to the middle of my departed lover's name. Yet in terms of our personal history, it turned out to be prophetic: it was written during the middle and not the evident end of our relationship.

Over the course of a year, I managed to complete a 331-page long manuscript. I was helped along by yet another disaster in what I soon came to recognize as my personal *annus horribilis.* In June of that year, while hiking through a creek bed on my ranch, where I had lived since my divorce from my second wife in 1976, I broke through some rotten timber and fell over backward—all of me, that is, except my left leg, which, the knee having been surgically fused some years before, failed to bend along with the rest of me: it remained lodged in the rotten log, immobilized and practically fragmented. The accident occurred in a very steep and rarely visited gulch on my 1200-acre ranch, some 30 minutes by car from Stanford University. If I had not been accompanied by a group of my graduate students, I surely would have died. As it was, it took seven hours and over a dozen professional rescuers to winch me up to the waiting four-wheel ambulance, and truly elephant-sized doses of morphine to blunt my pain. The post-operative recovery kept me in a cast for eight months, time I used to good effect, fueling my *magnum opus* of heartbreak and revenge with memories of quite another pain. The rest I finished on airplanes, in hotels, and at scientific conferences.

My ability to complete an entire novel under less than ideal conditions impressed me sufficiently that I actually started to look into getting it published. Fortunately, my muse intervened. On May 8, 1984, I received a note and flowers from Diane Middlebrook, whom I had not expected to ever encounter again. Her message, essentially, was: 'A year has passed. Let's talk.' Of course, I accepted; though instead of flowers I presented her in return with a selection of the more brutally frank chapters of 'Middles'.

Our reconciliation took some months, but even from the early stages of our rapprochement, we realized that we would never part again. In some silly and yet fundamental way, 'Middles' helped. Diane tried to persuade me that the novel was unpublishable on many grounds, discretion being only one of them. I promised never to publish that manuscript if we got married, and marry we did on 21 June 1985. And yet I hedged. I said I would not *publish* 'Middles', but I never promised not to cannibalize the manuscript.

Although in everyday life, I possess a sense of humor, 'Middles' suffered from a conspicuous lack of it. Its structure was too linear, the dialog forced. (I conceded that point, blaming it on four decades of still-too-ingrained scientific writing, with its monological and impersonal monotony). But as my severest critic pointed out, 'Middles' did show occasional flashes of insight. And most important, it demonstrated that I possessed a writer's discipline. 'If you want to write fiction,' my new wife advised, 'try short stories.'

Short stories, I soon found, are the ideal vehicle for an autodidact: one learns concision, and becomes cold-blooded enough to kill one's literary darlings—the *irresistible* bits of trivia, nuggets of pure joy, that make their way so easily into my conversation, but never seemed to find breathing room in the confines of my stories. Most important for the scientist who hardly ever revised his scientific manuscripts, one acquires the habit of repeated revision. Scientific writing aspires to the pure transmission of information: there content is king, style counts for little.

The first excerpt I extracted from 'Middles' as a vehicle for a subsequent short story was almost totally autobiographical. *What's Tatiana Troyanos Doing in Spartacus's Tent?* is my shortest short story and longest title. Eventually it appeared in *Cosmopolitan* (UK) magazine as well as in German and Italian translations. Here are the beginning paragraphs:

Through years of sexual reality, I dreamed ever so often of a woman lover who'd sing while coupling with her man. Once a husky-voiced woman, who'd brought her guitar, started to sing in a stunning contralto. Lying on my bed, exhausted and content, gazing at the naked woman strumming her instrument, I was about to ask her whether she could . . . But then I chickened out; I was afraid she'd just laugh.

Years later, I happened to go to a performance of Monteverdi's *L'incoronazione di Poppea*, set in the time of Emperor Nero before he went mad, with Tatiana Troyanos singing the lead role. About half-way through the love-scene on the couch between young Nero and Poppea, the performance assumed such erotic overtones that I began to squirm in my seat. I don't go to operas for sexual titillation; except for an occasional *Salome* or *Lulu*, it's the music that excites me. But this was different. Suddenly I realized that Troyanos was the woman of my fantasies who'd walked into Spartacus's tent over two thousand years ago.

I was a relatively late bloomer, a virgin until nearly twenty. But as an otherwise precocious teenager, I made up for it soaking in the delicious warmth of a full bath. Not in one of those modern tubs . . . No my passions throve in a real tub—one of those huge pre-war jobs—where I'd float, water up to the chin . . . incandescently copulating with Veronica Thwale . . . Veronica, in her twenties, was the deftly cunning courtesan I'd been waiting for and had finally found in Aldous Huxley's *Time Must Have a Stop*. God, she was something! Once our passion slipped me forward in that six-foot tub, so that I choked on soapy water.

But the spur to my longest and most lucid dream was *The Gladiators*—Arthur Koestler's version of the slave uprising led by Spartacus. Don't get me wrong: as time passed, as I became a

> man, there were months, even whole years, when Spartacus did not exist. But the vision never departed totally from my memory. When I had a lover whose climax always ended in such a long-drawn cry that we could never meet in a hotel, I wondered more than once how Spartacus had handled this in his tent on the plains of Campania. And when I saw Spartacus at the Bolshoi in Moscow thirty-five years after I'd read Koestler's novel, I felt a pleasure reawakening . . .

I then dropped 'Middles', though only temporarily, as a source of inspiration to turn to *real* fiction as I thought of it then, not fully realizing that all fiction writers are consciously or unconsciously autobiographers wearing masks. My second short story, initially called 'Cohen's Dilemma', was my first entry in what has turned to out be the genre for most of my novels: 'science-in-fiction'—not to be confused with science fiction—in which I describe the tribal culture of contemporary science. In 'Cohen's Dilemma' I wrote about a scientist given the Nobel Prize for work that turns out to be flawed—a story that pleases me to this day because of its plausible premise and its ambiguous end. It shows a scientific superstar falling in love with a spectacular hypothesis (how many of us have not yielded at one time to such dangerous temptation?) and then convincing his favorite collaborator that the experiment, designed to confirm the validity of his hypothesis, *had to work.* But when the lab's Prince of Wales delivered the hoped-for experimental goods and the work was published, a key competitor could not reproduce them. Was the hypothesis false? Was the collaborator's work in error? Or was it . . .?

'Cohen's Dilemma' met the usual fate of a first submission: a printed rejection slip from *The New Yorker.* But the next attempt succeeded: Fred Morgan of the *Hudson Review* accepted what became my first published short story. Morgan requested that I change the name Cohen on the advice of an editorial consultant of Nobel Prize stature, who felt that Cohen was too common a name for a fictional Nobelist. Rather than arguing, I proposed 'Cantor,' since for reasons obvious to any reader of my story, the name had to start with a C; for personal reasons, it also had to have a Jewish connotation. But the editorial consultant was too well informed. 'What about Charlie Cantor?' he inquired, referring to the then-chairman of the genetics department of Columbia's College of Physicians and Surgeons. I was so anxious to see my first short story in print, that instead of arguing that Columbia's Cantor surely carried no trademark on that name, I offered an 's' for the 'n.' *Castor's Dilemma* proved to be the vehicle that eventually converted me into a novelist.

Chapter 7

Science-in-fiction is not science fiction. Is it autobiography?

I

The first half of 1985 started auspiciously: Diane and I were married on my son's birthday, 21 June. The ceremony was followed by a glorious wedding feast at my ranch, attended by many of our Stanford University students and colleagues. Its artistic high point was the commissioned music for three flutes and soprano by John Adams set to lines by Wallace Stevens and the wedding dance created by Rhonda Martyn with her troupe of dancers. My scientific research at Stanford, with nearly two dozen graduate and postdoctoral collaborators, was going full steam. My health had recovered so completely from the accident and enforced rest that I felt ready to embark on a trek from Tibet to Nepal. It was not the first time I had embarked on such an expedition—four years earlier, in western Bhutan, a fabulous trek to Chomo Lhari, Bhutan's highest and holiest Himalayan peaks, had demonstrated that my fused left knee could cope with such rugged terrain. I felt buoyant, rejuvenated, and ready to take on the Himalayas themselves.

But not to the point of foolhardiness. Ordinary caution suggested that, at my age, a thorough physical examination should be a part of my preparation for the high altitude trip. Instead of the clean bill of health I anticipated, I was diagnosed with colon cancer. Instead of breathing Himalayan air, I inhaled an anesthetic for over five hours. The operation went overtime when my ureter was cut by mistake—a team of urologic surgeons had to be called in to make the necessary repairs. Barely a new bridegroom, I suddenly came face to face with my mortality. Would I survive beyond the next few years? What if I had known five years earlier that I was harboring a cancer? Would I have led a different five years? Of course, I thought: and so what about the next five years? The question, which became more pressing as the auguries from the pathologist had still not arrived, led me to toy with the idea of embarking on one more intellectual life, one far different from the preceding 50 years.

I still had a large research group, and many doctoral candidates who had several years of research ahead of them before they would be granted their degrees. It was clear that I could not abandon them, nor was I prepared to drop instantaneously an exciting line of research.

Once out of the hospital, however, I took the first step: without telling anyone, I decided not to take on any more graduate students. I did accept new postdocs, because they only came for a year or two. And I was not sure yet whether a literary career was actually feasible. I was (and still am) too ambitious to consider the act of writing by itself sufficient: I had been a publishing creature far too long to imagine an audience of one an adequate validation. I was also quite aware of how much of the craft I had still to learn. But how to go about it? I don't know how long I would have cast about, had not my muse intervened once more.

By the summer of 1986, I was fully recovered from my surgery, and Diane agreed to serve as director of a Stanford summer program at Oxford University. For the first time in my life, I became a faculty spouse. Every day I would drive her to the Stanford site at Magdalen College, take my morning swim, buy some groceries, and then settle down for the rest of the day to write. We were staying in a two-story house on Woodstock Road, the guests of my son's in-laws, Robert and Betty Maxwell, who in addition to running Pergamon Press owned Headington Hill Hall, the estate where I took my morning swim. The upstairs occupant was another recipient of Maxwell hospitality, the former Prime Minister Harold Wilson. One afternoon, he dropped in to regale us for about an hour with recollections of prime-ministerial minutiae that my wife found maudlin. The Anglophile in me was charmed.

I had decided to follow Diane's counsel to concentrate on short stories. After the completion of *Castor's Dilemma* and my *Tatiana Troyanos* story, I had started drafts of three other stories. I revised and polished them and then continued without interruption in a euphoria of fantasizing, discovering the giddy freedom of the imagination on the loose. This is how I write, and what I value most about the writing: I know how I want to begin; I can visualize my main characters; I even have a good inkling of how to end the story. Yet each time, my protagonists change 'their' plans; they introduce me to other persons, previously unknown to me; they take me on excursions I never anticipated.

These fishing expeditions into my unconscious started to teach me a great deal about myself, and of course this fascinated me. Scientists are on the whole not self-reflective. I certainly never was. But as I started to examine what made me tick, my stories started to assume a psychoanalytical resonance.

'I still think it's absurd. There I was in the hospital, my arm in a cast, feeling sorry for myself, and still not knowing precisely how I'd smashed up Bea's car.' Those were the first two sentences of 'The Toyota Cantos', a story I started to write in 1985 while flying to Italy to present a lecture before an Italian Gynecological Congress in Salsamaggiore. The Italian connection was important, because I had made my hero a Dante specialist from New York University, and his wife a creator of bumper stickers. Her stickers, seemingly a whimsical sideline, the kind of thing a bored faculty spouse might undertake to occupy her time, were peculiar in that they used for their texts passages from *The Divine Comedy*. And that they functioned as a code, a covert, failed attempt to communicate with her husband. I hadn't read Dante for some decades, but now I spent my otherwise unoccupied hours scanning every translation I could get my hands on: Cary's, Binyon's, Singleton's, Sisson's, Carlysle-Wicksteed's, Mandelbaum's, Musa's, and Anderson's. Only when I hit upon John Ciardi's colloquial English version did I settle down to construct the list of Dantean bumper stickers that became the centerpiece of my story. Overcome with the romance of this first excursion into true belles-lettres, I stopped, after my scientific lectures and a visit to the Gori Collection outside Florence, for a couple of days in Lérici, where Byron, Shelley and their friends cavorted and where Shelley ultimately drowned, there to weave Dante and bumper stickers into the fabric of 'The Toyota Cantos'. Sitting on the balcony of a *pensione* overlooking the Bay of Genoa, I explored the inability of a sophisticated, long-married, academic couple to communicate on deeply personal issues. I needed no shrink to tell me why that subject had worked its way into my favorite short story. During that Oxford summer I completed a collection of eleven stories, which ended up as my first published book of fiction under the title *The Futurist and Other Stories.*

II

The following year, Diane and I decided to repeat a concentrated summer of writing, but this time in London, where she was to work on her biography of Anne Sexton (eventually a *New York Times* bestseller), and I on an expansion of *Castor's Dilemma* into a full-length novel. Within weeks of its appearance in the *Hudson Review,* I had started to receive letters about 'Castor': 'Had the postdoc cheated?', 'Was Professor Castor thinking of suicide?' and 'Do you really find it necessary to wash dirty lab coats in public?' This last response is one I continue to hear from time to time, and my response is always the same: People working in white coats are bound to get dirty—a fact of life that white-coated scientists can not afford to ignore. But as long I was going to be doing the washing in public, it behooved me to be careful myself, so before submitting the manuscript to a publisher, I headed for the Royal Society Library to confirm the non-existence of any tumor cell biologist named 'Castor.' To my chagrin, I discovered that such a biologist did indeed exist in Pennsylvania, whereupon I decided without further dilly-dallying to revert to the name I had originally chosen, Cantor. *Cantor's Dilemma* was the title under which Doubleday published the hardcover edition, and while the novel was reviewed very widely, both in the lay and scientific press, neither Charles Cantor of Columbia University nor any other Cantor ever complained to me. (A couple of years later, Charles Cantor and I were successive plenary lecturers at a scientific congress. The organizers decided to announce my talk as 'Cantor's Dilemma' and his as 'Djerassi's Dilemma.' The audience enjoyed it and so did we.) Some readers of the novel assumed that the initials C.D. of the title were a masked reference to my own name, but if this was the case, it was disguised so completely that it had escaped my attention.

The overall reception of the novel left little to be desired. The reviewer in the Sunday *New York Times* wrote, 'This novel's rendering of the scientific establishment is so precise that anyone considering a career in science should be required to read it.' I suppose most scientists hope to write a required text. I had never anticipated achieving this distinction in *fiction*, but the reviewer's recommendations were eventually adopted by a good many American colleges and universities—an audience that has demanded one or more reprints each year since the paperback edition was released by Penguin in 1991. Had I known all this, I might have been more philosophical about the first weeks following the official release by Doubleday in early October 1989. They were hardly auspicious.

The inaugural public reading was scheduled for 8:00 p.m. of 17 October at Cody's in Berkeley, a Bay Area bookstore, and seemingly the ideal launching pad for a local writer's career. My wife and I crossed the Bay Bridge shortly before 5:00 p.m. in order to have a leisurely dinner in Berkeley within walking distance of Cody's. I was about to feed a parking meter when the ground around us began shaking. Instantly, the air was filled with the hooting of car alarms, every Audi and BMW on the block hysterically heralding the apocalypse. We held on to the parking meters until the pavement stopped moving, and looked around us to find the city changed. We had just experienced the Loma Prieta earthquake of 17 October 1989. Across the Bay, back in the direction we had just come, a portion of the Bay Bridge had collapsed, and with it a stretch of elevated freeway. People lay dying under tons of concrete.

There was no reading that night. Instead, it took us over four hours in horrendous traffic and virtual pitch darkness to get back across the Bay, where we found our home, to our relief, undamaged except for one broken pre-Columbian pottery figure and books strewn all over the floor.

My next reading had been scheduled at the Palo Alto bookstore Printers Inc, but it was canceled: the building needed retrofitting to meet quake-safety standards. Meanwhile, on the other side of the continent, my publisher, Doubleday, was suffering seismic activity of its own, an upheaval which included the sudden departure of both the editor and the publicist to whom my book had been assigned. When the unexpectedly wide review coverage of *Cantor's Dilemma* propelled it to the Bay Area bestseller list, Doubleday was caught unprepared. The modest first print run was quickly sold out and no one pursued a timely second printing until after Christmas, thus making *Cantor's Dilemma* a short-lived collector's item for the two busiest weeks prior to Christmas. The post-Christmas reprint by Doubleday, though personally gratifying, hardly made up for this marketing bungle. If I were superstitious, I would have concluded that some powerful god of fiction was advising me to return to chemistry.

Exactly ten days after the earthquake, my son Dale and I took off for a round-the-world trip in fulfillment of an earlier professional commitment to arrive on 29 October in Male, the capital of the Maldives in the Indian Ocean. For reasons known only to the gods of air travel (easily as mysterious as the deities in charge of fiction), for that particular date the most direct route from San Francisco was to fly TWA to Vienna via New York, in order to catch, a few hours later, a twice-weekly Vienna-to-Singapore flight that refueled in Male. Our stopover in Vienna, though it lasted only a few hours, carried enormous psychic resonances.

We only had a few hours to kill between planes, which we decided to spend walking around the inner city. Our taxi passed the apartment house where I lived 51 years ago, and from whose balcony I saw the Nazi Brownshirts cross over the Aspernbrücke on the way to the predominantly Jewish quarter of Vienna. We got off near St Stephen's cathedral and headed along Kärntnerstrasse—Vienna's most elegant shopping street which leads to the Opera and is now closed to all vehicular traffic.

The weather was balmy that late October 1989, and many people sat in outdoor cafés or strolled in typical late Saturday afternoon fashion. Suddenly we came upon two musicians, a violinist and a cellist, who were surrounded by quiet, almost reverential onlookers. With a flourish of his bow, the violinist started a highly professional rendition of Mozart's *Eine kleine Nachtmusik*. The choice was so typically Viennese as to defy an attribution of kitschy or banal. As we slowed down to listen, I sang into Dale's ear in an undertone, 'Wien, Wien, nur du allein.'

'You sound like Grandpa at the opera,' my son whispered. I knew what he meant: at intermission on the way to the foyer, my father used to hum his favorite aria from the last act with a beatific smile. 'But what does it mean?' Dale asked, not knowing any German.

'"Vienna, Vienna, only you alone." Every Viennese knows this from birth.'

Dale squeezed my arm to let me know he understood this sudden outburst of Vienna-bred sentimentality. But then I saw three tall policemen—young and not yet pot-bellied—strut toward the circle of listeners, and my emotional house of cards collapsed. 'Watch it!' I warned my son with typical Waldheim-induced suspicion. 'They'll stop them.' 'No way', whispered my American son, who is not burdened by a Viennese childhood dating from the 1930s, which included chases by the grandfathers of these young cops, in identical green uniforms, when we played soccer in the park.

'Break up Mozart in Vienna?' Dale added incredulously, reaching into his pocket to produce one of Vienna's favorite chocolates, bearing Mozart's portrait on the wrapper. But I was right: The trio of police walked straight through the listeners' cordon, which opened like a cell being penetrated by an invading virus. The tallest of the cops stalked toward the violinist, getting so close to him as to force him to glance up from the music stand. Mozart died instantly. The young violinist looked crestfallen as the policeman addressed him with an arrogant smirk; the only detail missing from the picture forming in my mind was the tapping of a swagger stick. The other player quickly grabbed his open cello case and started to pack his instrument. Only then did one of the silent spectators rush up to slip a 20-schilling note into the case just as its cover dropped—the sole sign of sympathy or protest among the entire crowd. The officious trio moved on, perhaps for ten meters, beyond the sullen group; now they stopped, turned around in slow motion, and stared to be sure that Mozart was not revived. Past overfed loden-clad burgers gorging themselves with *Schlag*-covered cakes and coffee, my son and I walked on, each in his own way struck dumb. Yet fairness demands that I admit to never again having encountered such a scene in numerous subsequent visits to the city of my birth where buskers now reign seemingly unmolested.

October 29 was not only my sixty-sixth birthday but also the day on which we were supposed to be picked up in Male by the Soviet oceanographic vessel *Akademik Oparin*, which was on a 4-month long expedition that had started in Vladivostok and was scheduled to end in Leningrad the following February. I had encountered the ship two months earlier during a visit to the Far-Eastern Siberian Branch of the Soviet Academy of Sciences in Vladivostok, the *Akademik Oparin*'s home port, and was so impressed by its diving facilities (including an on-board decompression chamber) and shipboard laboratories that I managed to inveigle an invitation to participate in a forthcoming collecting trip to the northern Maldives. Once that was accomplished, it was not difficult to convince my Russian hosts to let me bring my son, who had learned scuba diving with me and had since become an expert diver. Dale and I were the only non-Russians on board, and also the only non-Russian speakers. But by using English, German, or Spanish (several of the scientists and crew had worked in Cuba), we managed.

But there were other barriers to communication to be overcome. In the late 1980s, Vladivostok was still a closed city. Indeed as far as international fax communication was concerned, despite the advances of Gorbachev's *glasnost*, the entire Soviet Union was basically non-faxable. But not the *Akademik Oparin*, whose fax machine could be accessed directly from the States via the COMSAT satellite system. Just as we sat down to a welcome-aboard dinner in the captain's quarters, the first officer broke in excitedly, saying 'A fax for the professor.' Instead of an emergency message, it turned out to be a photocopy of the glowing lead review of *Cantor's Dilemma* from the 29 October *Los Angeles Times*, with 'Happy Birthday!' scribbled in my wife's handwriting on the top. 'You can't enjoy pleasures like this in chemistry,' I told my son.

Cantor's Dilemma produced another enjoyable but also much more consequential effect. Prior to the actual publication of the German translation (only the apostrophe had to disappear in the German title), the *Frankfurter Allgemeine Zeitung*, one of Germany's leading newspapers, serialized the novel in its entirety over a period of two months, thus virtually guaranteeing its commercial success. *Cantor's Dilemma* has appeared in seven languages, but its German rendition has proved far more important to me than even the English version. The latter had important intellectual and professional consequences for me—mostly, but not entirely all positive—but the German publication carried totally unexpected psychological resonances in terms of reminding me of my European roots.

III

Cantor's Dilemma landed on fertile ground—a fertility fostered, oddly enough by the Pill. In 1969, after publishing almost 700 scientific papers and half a dozen monographs intended solely for an audience of organic chemists, I wrote my first two papers for a broader audience. The second, published in 1970 in *Science*—hardly a popular mass medium, but nevertheless with a wider audience than I usually addressed—bore the title 'Birth Control after 1984'. The paper contained not a single chemical formula. Rather, it addressed an issue that had simply been brushed aside at that time: What political, legal, and economic steps are needed to create fundamentally new methods of birth control—as different from the Pill as the latter was from the diaphragm?

My punch line—and hence the title—was that it would take approximately 14 years, and that if the steps I enumerated were not implemented promptly, nothing fundamentally new would be created. That contentious article proved to be the most widely read of my bulging bibliography. More significantly, the responses it brought caused me to extend my perspective to the social aspects of birth control—an area to which I, as a chemist, had hitherto paid little attention. As I describe in another chapter, that broader inquiry not only affected dramatically my teaching, but a few years later prompted me to follow Diane's footsteps in writing a book for a lay audience through the *Portable Stanford* series. These books, authored by Stanford faculty and distributed first through the Stanford Alumni Association to its members and subsequently published by a trade publisher (W.W. Norton), were meant to illuminate intellectual topics for a curious non-specialist readership. The preceding volume of that series, Diane Middlebrook's *Worlds into Words: Understanding Modern Poems,* was written during the first few months of our cohabitation. Observing the process of literary gestation in such affectionate proximity, I embarked on my first foray into popular non-fiction writing.

I have always been proud of *The Politics of Contraception.* It was a provocative, tough, yet readily accessible book that raised many issues which either had been ignored or even discounted during a time when the Pill in particular and birth control in general were contentious issues for the public, the media, and legislators. For years, it was used as a textbook in a wide variety of courses, even though it was not written with that purpose in mind. And suddenly, nearly ten years later, my next book—the novel *Cantor's Dilemma*—experienced the same fate. That novel continues to be reprinted primarily because of its extensive use in a wide range of courses: ethics in research, science in literature, sociology of science, and the many variations of the science-society-and-technology theme that have become the mode in academe. When I realized the use to which *Cantor's*

Dilemma was being put, I decided on a literary gamble: to pursue fiction writing for didactic reasons. That this was a gamble I never doubted. Even as a scientist, accustomed to writing nothing but didactic prose, I knew that a novel labeled 'didactic' by book reviewers was likely destined for the remainder bins. Look up the word in *Webster's*, and the first thing you read is 'designed or intended to teach'—harmless enough: but listen to the literati say the word, and the pejorative overtone is clear. I solace myself (if I need any solace, given the way my gamble has paid off), by telling myself that I aim for the expanded definition in *Webster's* dictionary: 'intended to convey instruction and information, *as well as pleasure and entertainment* (emphasis added).'

But why take the gamble at all? It wasn't that I could do nothing else—despite a lifetime of utilitarian prose, I had found that I had no trouble spinning the stuff that dreams are made of. Rather, I had a very specific aim in mind, an attempt at addressing a problem that my recent leap from science to the arts had made pressingly clear and personal to me. The gulf between the sciences and the worlds of the humanities, the social sciences, and mass culture, is increasingly widening, yet scientists themselves spend precious little time attempting to communicate with these other cultures. To a large extent this is due to the scientist's obsession with peer approval. Moreover, his tribe offers few incentives to communicating with a broader public—and more than a little disincentive. One of the best and most serious of scientific popularizers, the late Carl Sagan, paid a heavy professional price for his popular success—including being blackballed from the National Academy of Sciences. I am now at an age where disapproval or indeed, even approval, by my scientific peers can do little for my scientific career or self-esteem. I certainly won't be thrown out of the National Academy of Sciences (to which I was elected 40 years ago) for writing fiction. Yet I feel strongly as a scientist and as a teacher that the scientific culture must be illuminated for a broader public that otherwise—largely out of scientific illiteracy—cares little for it. Why not try smuggling such concepts into the consciousness of a wider public in the guise of fiction?

And so I call the literary genre in which I work 'science-in-fiction.' It was important to me to differentiate what I do as clearly as possible from science fiction. For me, the most important difference is that in science-in-fiction all the science or idiosyncratic behavior of scientists described in it is plausible. None of these restrictions applies to science fiction. By no means am I suggesting that the scientific flights of fantasy in science fiction are inappropriate. But if one actually wants to use fiction to smuggle scientific facts into the consciousness of a scientifically illiterate public—and I do think that such smuggling is intellectually and socially beneficial—then it is crucial that the facts behind that science be described accurately. Otherwise, how will the scientifically uninformed reader distinguish between what science is presented for entertainment and what is informative?

But of all literary forms, why use fiction? In contrast to, say, history (a frequent subject of 'didactic' fiction), the majority of scientifically untrained persons are afraid of science. The moment they learn that some scientific facts are about to be sprung on them, they raise a mental shield. It is that portion of the public—the ascientific or even antiscientific reader—that I want to touch. Instead of starting with the aggressive preamble, 'let me tell you about my science,' I prefer to start with the more innocent 'let me tell you a story' and then incorporate real science and true-to-life scientists into the tale.

My ambitions, however, reach farther than just wanting to describe what *research* scientists perform. Good science journalists already fulfill that role, and frequently they do it very well. But to bridge the gap between C.P. Snow's two cultures, to make science as real to people as any other job a human being might do on an average day, it is also necessary to illustrate how scientists *behave*. And it is here that a scientist-turned-author can play a particularly important role.

Scientists operate within a tribal culture whose rules, mores and idiosyncrasies are generally not communicated through specific lectures or books, but rather are inculcated through a mentor-disciple relationship in a form of intellectual osmosis. We learn—and earn—our scientific 'street smarts' by observing and serving the mentor's self-interested concerns. As we struggle to support our mentors in their jousts with journals, the constant jockeying with colleagues and competitors over position and priorities, the order of the authors, the choice of the journal, the quest for the grail of academic tenure, then grantsmanship, *Schadenfreude*—even Nobel lust, that most exalted failing of the great—we learn how the game is played, and unconsciously we absorb the lesson that the game *must* be played. In other words, it's a career—one with its share of glass ceilings and close friendships, brutal competition and subtle ethical nuances—but a career all the same, something done by and for and with people. People in white coats, speaking an impenetrable jargon, but people all the same. To me—as a scientific tribesman for over four decades—it is important that the public does not look at scientists primarily as nerds, Frankensteins or Strangeloves. And because science-in-fiction deals not only with real science but with real scientists, I feel that a clansman can best describe a scientist's tribal culture and idiosyncratic behavior.

I would be lying if I claimed that I had spelled out all of that motivation at the very outset of my literary writing in the mid-1980s. These thoughts started to crystallize during the genesis of *Cantor's Dilemma*. They only grew into wider implementation as a result of the reception of that novel—a reception that encouraged me to lock the door of my laboratory from the outside and to throw away the key.

Chapter 8

Behind the scrim of fiction

I

As far as I can remember, the last novel I read in German was Arthur Koestler's *The Gladiators*; I was 18 or 19 years old at the time. Half a century passed until I read another novel in German, this one with an attention I had never before given a work of fiction. As I turned the pages of Ursula-Maria Mössner's translation of my own novel, *Cantor's Dilemma*, long-forgotten German idioms surfaced from some Viennese depth that I had assumed had long ago been covered by an impenetrable thicket of Americanisms.

In hour-long telephone conversations between San Francisco and Ulm, Ully Mössner's home in Southern Germany, I attempted to convince that sophisticated German translator to modify a phrase here or there, using—what I thought—would be a more precise phrasing of my original text. (Even the word 'English' is defined with unimpeachable precision in Germany:

all my books state on the cover, '*Aus dem Amerikanischen*'—'from the American'). But I grossly underestimated the stubborn professionalism and diplomacy of the person who has since become my verbal German *alter ego*. Ully Mössner's polite demurrers—'I'll think about that' or 'I'll let you know'—changed eventually into undisguised hilarity as she confessed that in most instances, my modifications were based on a Viennese of the 1930s. In short, she said, they dated me. I finally conceded to myself that in spite of my silver hair, my German had never grown out of its teens.

When the German translation came out, my German publisher arranged a reading tour, starting in Berlin. The practical challenges of reading a text aloud in the translated German of a modern German woman presented no difficulties, however, and my slight Austrian accent, now even scented by whiffs of occasional Americanisms, certainly did not put off the audiences. Initially, I was worried how I would cope with extemporaneous answers in German to *ad hoc* questions. But as I proceeded from Berlin to Hamburg, then Köln, Frankfurt, Braunschweig and onwards, my *Deutsch* loosened almost hourly. Even live TV and radio interviews ceased to concern me. After nine years, much of my German has now returned. On at least one occasion, I have even dreamed in German—to me the ultimate testimony that a reviving mother tongue is poking holes in a psychic barrier that, prior to the mid-1980s, I had assumed had set into impenetrable concrete.

It is strange that this recovery of my Central European roots did not occur in Austria, the country of my birth and early education, but rather in Germany. This was the country, after all that had severed those roots in the first place when Hitler's legions had driven me from Europe. All my relatively infrequent visits to Germany had been as an adult, to scientific lectures or congresses, and these had been held entirely in English. Except for a smattering of small talk with waiters and taxi drivers, and the occasional skimming of some German newspaper, I had always considered myself, and acted like, a visiting American. All that has changed since the publication of my fiction in Germany. It took a modern German voice, and a woman's voice at that, to bring me to terms with my European origins.

II

As I explained in the preceding chapter, the reception of my first novel, *Cantor's Dilemma*, prompted me to continue working on a projected trilogy in my 'science-in-fiction' mode. The seminal idea for the second volume, *The Bourbaki Gambit*, came to me at Princeton University, where I spoke in 1988 at the retirement party of my friend and close collaborator, Kurt Mislow—at that time arguably the most distinguished chemistry professor on the Princeton faculty. During the festivities, attended by over 100 of Mislow's former graduate students and postdoctoral fellows, as well as faculty members from at least a dozen other universities, my friend drew me aside to show me a curt one-sentence missive he had just received from the secretary of the Princeton board of trustees. In its coldness, its gracelessness, and in the succinct legalese in which it spelled out the terms under which the Princeton laboratories would tolerate Mislow's continued presence, it epitomized a kind of corporate ruthlessness that has become all too common in academic science. Clearly, it had grated on my friend.

It grated on me as well, perhaps because I thought that my own university was unlikely to be more graceful. And even though I was not even thinking of retirement, I wondered how I would respond if faced with a similar example of institutional coldness. This kernel of irritation grew eventually into a pearl of fictional revenge. Within months, I had created a plot in which four distinguished scientists, either legally retired or bribed into retirement, had formed a cabal to pool their resources for continuing research under the *nom de plume* of 'Dr Diana Skordylis.' I modeled their collaboration on that of a famous group of French mathematicians who actually have been publishing collectively for several decades, under the assumed name 'Nicolas Bourbaki.' Here is their plan as described in *The Bourbaki Gambit*:

> I think it would be great,' I began, 'if a group of older scientists could take on the cult of youth and teach it a lesson. Scientists who still have a lot on the ball, like . . .'
>
> I was trying to think of some names that would mean something to her, but she beat me to the draw.
>
> 'Like you?' she asked.
>
> Well, of course like me. I had intended to be modest and mention someone like Linus Pauling, but since she brought up my name, why not? 'Sure,' I said, 'for instance me. Now suppose some of us got together. Not many—half a dozen, maximum. We have already published a lot, received plenty of kudos' (though never enough of them, I thought), 'so a few more papers under our own names won't make much difference. Suppose, however . . .'

> 'Yes,' she broke in and grabbed my arm, 'why not form a Bourbaki of old scientists?' Seeing me cringe, she quickly corrected herself. 'You know what I mean, not really old, just . . .' She groped in vain for a word, and we both broke out laughing.

I even went farther. Except for the signature, I used a verbatim copy of the letter that had vexed Mislow during his retirement party.

> 'Of course you can continue your research, Max. How could you even ask? Our star in the biochemical firmament,' the Dean had said on the day we signed the generous agreement. 'You do understand that university rules preclude making long-term contractual commitments with emeritus professors. But you can trust me, Max. I'll get you something in writing.' So I trusted him, and eventually I did get something in writing: that infamous one-sentence letter from the secretary of the Board of Trustees—a missive that still makes my blood boil.
>
> '*Dear Sir*.' McLeod had written. Not *Dear Professor Weiss* or even *Dear Weiss*. My thirty-four years at Princeton—twenty-seven of them as Donohue Professor of Biochemistry—apparently counted for nothing. Just '*Dear Sir: I have the honor to inform you that at a meeting of the Trustees of Princeton University held today, you were appointed Senior Research Biochemist in your department, without stipend, for the period June 1, 1988, to May 31, 1989. Respectfully yours, Seth T. McLeod*.' It isn't often one remembers an entire letter verbatim, but this one, in its insolent brevity, was one for my permanent mental files. It wasn't so much the 'without stipend' caveat, although it could have been done with a dash of grace. But one year at a time?

> Despite the bitter taste left each time I inwardly recited this single sentence, I wasn't sure I wanted to forget it: now that I had been cut adrift, my resentment, at least, was something to hold onto.

What gave me particular pleasure, beyond my obvious identification with the age group of my fictitious protagonists, was to write the novel entirely in the first person singular. I do not believe that in over 40 years of research writing I ever used the word 'I.' This cultural shackle I had worn for so many years broke in an avalanche of 'I's in *The Bourbaki Gambit.*

I started the novel in London in 1989, and completed it within two years that included a five-week residency at the Rockefeller Foundation's Villa Serbelloni in Bellagio on the shores of Lake Como. The jacket cover of *The Bourbaki Gambit* features an image from Magritte's 'The Sorcerer,' depicting a man performing four different actions with four hands. I like to think that Magritte painted it for me, to represent my four fictional scientists at work under a single alias. Just as with my first novel, *Bourbaki* was reviewed widely in both the general press and scientific journals. Two excerpts, reproduced shamelessly herewith ('A subtle meditation on the scientific personality . . . as creative as any organic potpourri that Djerassi might have mixed up in his laboratory') and ('Probably the quintessential science novel of the past year') pleased me especially and not just because of their complimentary tone. Neither these two reviewers nor any others complained that all my principal characters had either passed or were approaching the biblical age of three score and ten years. The reason for favoring that age group was more compelling than just the personal identification I acknowledge freely.

In Japan, North America, and Western Europe, where most of the frontier research in biomedical science is currently conducted, we are witnessing the emergence of geriatric societies: not too far into the twenty-first century, a quarter of the population of these regions will be beyond the age of 60. We read all the time about the social problems associated with such a skewed age distribution—problems of medicine, politics, economics, and even recreation. Some other implications of the aging of our culture are discussed much less frequently; among these is the rapidly increasing size of the geriatric element of the intellectual elite. In the process of writing a semi-fictional account of the challenges these demographic changes raise in academia, I was also forced to face my own attitude toward 'retirement.' In view of my birth date, I do not fall within the group of academics subject to compulsory retirement, a fluke that tempts me to fantasize about becoming the first non-retired centenarian on the Stanford faculty. But what caused me to even dream about possibly turning into a Strom Thurmond of academia? As I wrote about the differing attitudes of my *Bourbaki* characters, I started to wonder about my own dismissive attitude toward voluntary retirement. Was is pride or just a manifestation of a workaholic's machismo? My own answer can be found in the very first chapter of *The Bourbaki Gambit*:

> A change in the woman's tone alerted me to an incoming question. My ears switched back to her wavelength: ' . . . with all that, why are you retiring?'

> 'It's best to retire when people ask "Why are you retiring"?' I replied, 'and before they ask "When are you retiring"?'

The inability of scientists to write under a *nom de plume* has always intrigued me. We all know that the desire for name recognition has to do with individual identity—with ego, the ultimate source of all this hunger for success. So why can some writers but no scientists cope with anonymity? Because for the former, reputation is ultimately established by general acclaim, whereas a scientist's standing *as scientist* is never made by the general public. Here is what Max Weiss, the main character in *The Bourbaki Gambit*, had to say on that topic:

> There is one character trait, which is an intrinsic part of a scientist's culture, and which the public image doesn't often include: his extreme egocentricity, expressed chiefly in his overmastering desire for recognition by his peers. No other recognition matters. And that recognition comes in only one way. It doesn't really matter who you are or whom you know. You may not even know those other scientists personally, but *they* know *you* through your publications.

But writing about the idiosyncratic features of a scientist's life was not enough for me. I also wanted to include some important science and the plot of *The Bourbaki Gambit* offered me a seamless way of accomplishing that. Here is a description of my aging scientist's plan to work as a collective to revenge themselves against the persisting cult of youth in an increasingly geriatric society:

> We'll put our heads together and go to work on some real intellectual gambles. The kind of thing most younger scientists

> aren't willing to try because they may risk tenure or other advancement. If one of these works out, it will cause a sensation. Everyone will want to meet that new star, and then they'll find out . . .

The research project my *Bourbaki* scientists picked turned into the discovery of PCR—an acronym standing for polymerase chain reaction—one of the most important methodological advances in the biomedical sciences. The acronym was not even coined until 1986, but within three years of its invention (by scientists at Cetus Corporation in California), PCR had swept the biomedical field and was recognized in 1993 through a Nobel Prize in Chemistry. Even the *Jurassic Park* fantasy would have lost whatever plausibility it may possess without the existence of PCR. Yet how many laypersons have heard of PCR? And of those who have, how many can explain the concept behind it? I dare say that if the 12 jurors in the O.J. Simpson trial had understood it (or perhaps had just read chapter 19 of my novel) the outcome of that case might have been very different.

III

Whatever the literary qualities of these first two novels, I was pleased with the didactic achievement—the large amount of insider information I had been able to report. I felt I had managed to present an accurate and wide-ranging picture of the arcane rituals of academic scientific research. That may well be the reason why *The Bourbaki Gambit* has now joined *Cantor's Dilemma* as recommended reading in college courses of the 'science, technology and society' variety. Having accomplished as much, I chose to focus next on a particular question that had come out of my own time in science, one I have been asked ever since the Pill gained popular acceptance: 'Why is there no Pill for Men'?

There is no single, all-encompassing answer to this seemingly simple question, because it is accompanied by a lot of hidden baggage, way beyond straightforward scientific or even commercial considerations. It took me two volumes to get that out of my system, which explains why my intended trilogy of 'science-in-fiction' grew into a tetralogy. In these last two installments I tackled the question why no such male contraceptive will be available for at least the next decade or two. *Menachem's Seed* and *NO* focus on those areas of male reproductive biology that are currently high on the list of research and development priorities of research scientists and the pharmaceutical industry. These involve the treatment of male infertility—the subject of *Menachem's Seed*—and of male impotence—as covered in detail in *NO*—but not male contraception.

> 'I knew it!' the woman exclaimed triumphantly. 'All female reproductive biology means to you is contraception. But when you men work on your own sexual apparatus, all you worry about—'

The 'woman' in the preceding paragraph could well have been the late Margaret Mead (whom I quoted along these lines in an earlier chapter) or many other like-minded critics. Instead, she is one of the characters in *Menachem's Seed*, where women's understandable concerns are one of the components in a fairly complex plot. But to balance the discussion, I made the protagonist of *NO*, who discovers a treatment for erectile dysfunction, a woman.

In *Menachem's Seed*, I also entered the arena of international policy, giving a semi-fictional account of the Pugwash Conferences on Science and World Affairs. Even though it had been around since the late 1950s, Pugwash was essentially unknown to the general public, until 1995, when it received the Nobel Peace Prize. The publicity surrounding that award motivated me to draw on my own experiences with the Conferences from the mid-60s to the early 1980s. Much of the action in *Menachem's Seed* is set at a semi-fictional Conference on Science and World Affairs in Austria that resembled the actual Pugwash Conference where some of the first unofficial contacts between Israelis and Palestinians occurred—contacts that were officially denied and certainly could not have been undertaken through ordinary diplomatic channels. The Menachem of the book's title is based in part on a real person who was intimately involved with Israel's atomic policy. In my novel, he is also the romantic interest of an American scientist whom I used to illuminate the problems of modern professional women struggling with the conflicting priorities between their biological clock and professional ambitions.

But the international politics described in *Menachem's Seed* was only an excursion from my ongoing interest in the different areas in which science is practiced. In *NO*, the fourth volume of my tetralogy, I turned my authorial attention to a sphere where the behavioral and even cultural patterns differ from those in academic science, namely industry. Since I had straddled the academic and industrial worlds for decades, I used that experience extensively to focus in *NO* on the small, entrepreneurial, research-driven enterprises sometimes collectively referred to as 'biotech.' In contrast to 'big' industry, the biotech companies of the 1980s and 1990s are a uniquely American phenomenon born out of academia—much of it right in my own backyard, the San Francisco Bay Area. Because of their intellectual origins in educational institutions, these biotech ventures have also generated a series of contentious problems, arising from the interaction of profit-driven enterprises with supposedly non-profit institutions and (ideally) disinterested individual scientists. These have caused numerous legal, philosophical, and ethical debates that will continue to influence the conduct of science within the academy, as well as the ways scientific information is disseminated into the economy and culture at large. To the extent that such topics are worth teaching, my fictitious *NO* can easily serve as a realistic casebook text of such contentious issues.

While a two-letter title would seem to leave little scope for ambiguity, the meaning of my *NO* is complicated. It refers, of course, to the many uses of the negative expletive, but it also happens to be the chemical formula of nitric oxide, an industrial gas and global, environmental pollutant. Yet some of the hottest recent biomedical research (recognized by the 1998 Nobel Prize in Medicine) has also shown that NO fulfills a singularly complicated and sophisticated function in the human body, where it serves as a biological messenger, indispensable in a staggering variety of processes, including penile erection. This, in turn, led me to use in my novel the therapeutic treatment of male functional impotence—the research behind drugs of the Viagra type—as the vehicle for illustrating the role of a 'biotech' company in contemporary biomedical research. In the process,

I managed to bring all the main figures from my preceding three novels together and thus complete the tetralogy in a manner not unlike that in which C.P. Snow used characters from his earlier books in his own 'academia-in-fiction' series.

I did not pick research on male impotence—or erectile dysfunction as it is now called more diplomatically—for prurient reasons. Rather I wanted to illuminate an area of male reproductive biology that is now receiving heavy attention, in marked contrast to the dearth of interest in male contraception; at the same time, I wanted to explain the historical path that led to the understanding of the role of nitric oxide (NO) in penile erection.

> 'I kept it simple. I just told them not to confuse nit*ric* oxide, NO, with nit*rous* oxide, N_2O, which is ordinary laughing gas. But I couldn't just leave it at that, could I?'
>
> Shelly shook her head sympathetically.
>
> 'I suppose that's where I slipped. My instincts as a pedagogue took over. I told them that these tiny wafts of NO mediate an extraordinary range of biological properties ranging from the destruction of tumor cells to . . .' he slowed to a halt, and gave a rueful smile. 'I *could* have said 'to the control of blood pressure' and let it go at that. But instead, I said, 'to causing penile erection.' But note: the words 'stiff pricks' never crossed my professorial lips!'
>
> 'Felix!'
>
> 'Do I perceive a touch of disapproval in my spouse's eyes?'
>
> The hard edge behind the banter did not escape his wife. 'Disapproval? Certainly not. Of course, you could've been a bit more diplomatic and not displayed your penile erection—'
>
> 'Shelly! Puh-leeze.'

She raised her hand. 'You're too impatient tonight. Just let me finish. I meant that instead of flashing your penile erection work without warning, you could've prepared your audience.'

'Very clever.' Frankenthaler made no attempt to hide his sarcasm. 'In what zippered fashion would *you* have done it?'

'Oh,' she waved her hand airily, 'I would have started with mention of the *retractive* penis muscle of the bull. Frankly, the word 'retractive' is less aggressive—admittedly also less exciting—than what you exposed at dinner.' She leaned over to pat her husband's hand. 'I would have outlined Gillespie's work on nanc nerves and his interest in neural transmission, then mentioned that the only reason he picked the muscle in a bull's penis is that it is a particularly rich source of such nerves.'

'I'll be damned,' Frankenthaler exhaled. He meant it admiringly and Shelly, who knew her husband's nuances, took it as a compliment. Frankenthaler was amazed that his wife—a confirmed non-scientist—had remembered what he had told her about the origin of his own interest in NO research.

In a way, it's too bad I didn't finish that last novel a bit later, because the science behind NO is moving from penile erection to fertilization. In the year 2000, my Stanford colleague David Epel and his associates have finally found the answer to the century-long question what in the first contact between sperm and egg actually triggers embryonic development. It appears that it is none other than the release of NO by the sperm that prompts the release of calcium inside the egg—a phenomenon that Epel had shown earlier to be the factor sparking development in the fertilized egg. When my Felix Frankenthaler character in *NO* is accused of typical male bias in focusing on male potency in his work, he could now have used Epel's findings as a more polite justification than the following angry retort he gave in my novel to a woman questioning him at a dinner:

> 'Now wait a moment.' By now Frankenthaler didn't give a damn that he was supposed to be buttering up prospective donors. 'If you can't get it *up*,' he hissed, 'you can't get it *in*. Only then do we start worrying about birth control. And for your information, the person in my lab doing the work is a woman!'

IV

These days, many of my public lectures carry the generic title 'Science-in-fiction is not Science Fiction. Is it Autobiography?' The answer—in my case, certainly—has been affirmative. Only by dipping deeply and frequently into the well of my accumulated experience as a scientist can I even dare to attempt the task of explicating my tribe's mores to the world at large. Yet my admission that much of my work is autobiographical does not mean that the plots are biographical. They may be psychologically 'true,' but the bulk of the events they describe is fiction. And the part that isn't fiction is generally well disguised—with one exception: the novel *Marx, deceased*, which is not part of my 'science-in-fiction' tetralogy.

The promise I made in my marriage contract with Diane, that my first novel, 'Middles,' should go forever unpublished, did not include a guarantee that I would expunge it from my brain. During the intervening years, some aliquots of that suppressed work found their way into two of my short stories, *First-class Nun* and *Maskenfreiheit*. The rest simmered for nearly a decade, until it surfaced in a totally different context in *Marx, deceased* —a novel seemingly unrelated to my *'science-in-fiction'* tetralogy. I deliberately say '*seemingly* unrelated,' but of course it does bear some relationship to my tetralogy.

I wrote *Marx, deceased* after *The Bourbaki Gambit*. Part of the subject of that earlier novel had been the question of scientists' inability to write under a *nom de plume*. But what about a writer? I had long been interested in the function of celebrity in the world of science; now my early experiences as a writer of fiction led me to consider that question in the context of art. If one's central purpose is to win general acclaim, does it matter what name is ultimately acclaimed? One thinks of such writers as Eric Blair and David Cornwell, Mary Ann Evans or William Sidney Porter—did they care if their audiences knew them as George Orwell and John Le Carré, George Eliot and O. Henry? When one becomes famous, what is the relationship between one's reputation and one's real identity and worth?

Such questions ultimately lead the hero of my novel, Stephen Marx, to wish to read his own obituary, to be a fly on the wall and observe what people really think about him. Such a dream is surely shared by other ambitious and active fantasizers, including me. But Stephen Marx, a well-known New York writer, is not content merely to dream: he actually stages his death in a sailing accident. The obsequies over, the obituaries written, the question that Marx must face then is: what next? Holing up in San Francisco under the genderless new name, D. Mann, he writes and then publishes his fourteenth novel, But it is also his first, of course, under his *nom de plume*. As perceptive readers may already have anticipated, the title of D. Mann's first novel is *Middles*.

Here is a passage from *Marx, deceased*: a portion of D. Mann's *Middles* that I directly cannibalized from the manuscript I promised never to publish. My partial breach of that promise is as autobiographical as is the following excerpt.

> 'I know practically nothing about you other than that you are a scientist. Why don't you tell me something about yourself?'
>
> 'All right,' replied Nicholas. 'I'm a scientist, divorced, one grown-up child; but what is occupying me right now is a novel.'
>
> 'What's it called? By whom?' asked Gerald Bogen.
>
> 'I haven't yet picked a title.'
>
> 'You mean *you*'re writing a novel?'
>
> Nicholas wasn't sure whether admiration or amazement was behind Gerald's question. He was hoping for both.
>
> 'It's my first attempt.'
>
> Gerald had turned completely towards Nicholas, who could almost feel the searching, speculative—and, yes, admiring—look. 'What made you do it?'
>
> Nicholas decided this was the time to be honest. Not completely open, but honest. He would tell him why he wanted to publish it, not why he had written it. 'It's the competitive scientist in me. I don't know a serious scientist who doesn't want to publish his work. This is one of the major differences between us and the literati. Take a poet like Emily Dickinson. She hardly saw a single poem published in her lifetime. Yet she wrote superb poetry and got better all the time. It's unlikely that this could have happened to a scientist, who must be nourished by peer-approved publication. A male Emily Dickinson in physics or chemistry is inconceivable.'
>
> Bogen continued to look at him. 'Sounds to me like baloney, Nick. It's too simple. There must be more to it.'
>
> Nicholas felt himself bridling slightly under the charge—until he remembered it was true. 'Of course there is. It's the pleasure associated with literary writing: that's what I'm after. It's different

from the pleasure of scientific publishing. I'm quite well known in my field, primarily through my papers. But style in scientific publications really doesn't count; all the reader cares for is content, and all the editor looks for is concision. Most scientific papers are read only once, whereupon they either become part of the reader's data base or are discarded as excess baggage.

'Now that I'm writing a novel, I've started to read in a new way. I'm rereading books by authors who are real craftsmen. I already know the story, so I can focus on the hidden linguistic gems. They aren't really hidden; they are right there in front of us, but by attending primarily to the plot one misses a lot. These gems are exquisite candy; when I find one of them, I don't bite into it: I let it dissolve slowly on my tongue. I test it to see how long I can detect the aftertaste. Take Saul Bellow's *Herzog*. Have you read it?'

'No.'

'Bellow is a real craftsman. It's strange—as I reread *Herzog* I read slower and slower, until it took me several minutes to complete one page. I stopped at each sentence; I tested it; I asked myself, 'How did the man think of this phrase? How can I do it without copying him?' At one point, Bellow describes the look in Herzog's wife as 'terrifying menstrual ice.' Not bad, eh?'

'I doubt whether a woman would find it very apt.'

Nicholas snorted. 'I was rather taken by it. The trouble with Bellow's metaphors is that when I find one I fancy, I want to produce one myself. I got so enamored of his endocrinological simile, all I could think of was some glandular equivalent like 'dripping adrenals."

'That's as bad as "menstrual ice."'

Nicholas looked annoyed. 'I think you're being too literal, but I was unhappy with it for a different reason. It sounded too

> much like "adrenal rage," which I'm sure has been used lots of times, but I'd never heard of 'menstrual ice.' The beauty of such a combination is that, even though you've never encountered it, it makes immediate sense.'

If that is not biographical enough, let me conclude by mentioning a millstone I have been carrying around all my life: spelling my family name. The first two letters of Djerassi invariably invite confusion or insertion of an apostrophe, which I reject angrily as an unwelcome mutilation. So I start aggressively, without pause, to spell my name: 'D as in David, J as in Joseph, E as in . . .' at which point I am usually interrupted with the curt comment, 'just spell your last name!' No wonder that Nicholas Kahnweiler had this to say in *Marx, deceased*—another cannibalized morsel from the never-to-be-published 'Middles:'

> The most common question is, 'How do you spell your name?' The very first time I was asked this, shortly after my arrival in the United States, I, the sixteen-year old Nicholas, got flustered. When I tried to spell out my name the way I had heard others in America spell theirs with reference to geographical or personal names, all the words starting with a 'K' in my native tongue that came quickly to mind were 'Kalifornien' and 'Kairo,' which in English start with a 'C.' Kalamazoo was as yet unknown to me—I, the Central-European urban kid, would probably not have believed that a place could be called Kalamazoo—and neither Kansas nor Kentucky was as yet part of my spelling subconscious. Offhand, I couldn't think of a single English word starting with 'K,' so in desperation I came up with 'Kitsch,' and I have used this ever since. I got again stuck with 'W,' which of course I pronounced as a 'V.' The damn German confusion

between 'V' and 'W' got in my way, until I found an all-American word, 'Washington,' which even though I pronounced 'Vashington,' everyone knew what I meant. Kitsch and Washington stuck with me for the next 40 years. A psychologist would probably have a field day with the combination Kahnweiler–Kitsch–Washington.'

Chapter 9

A softer chemist

Until 1969, I would have described myself as a 'hard' scientist, the proudly macho adjective employed by chemists and other physical scientists to distinguish their work from the 'soft,' fuzzy fields such as sociology or even psychology. Next to physics, chemistry is the hardest of the hard sciences, the rock on which the biomedical, environmental, and material sciences all build their molecular edifices. Chemistry—the molecular science *par excellence*—also tends to pride itself on the social corollary of its flinty strength: it is the most insular of the hard sciences. In academia, we chemists are often the most conservative, refusing to climb beyond our self-imposed disciplinary walls. With the possible exception of engineers, we may even be politically and socially the most conservative of technologists. Given such an unyielding nature, it should not be surprising that, although we feel continually defensive in this time of rampant chemophobia, most of us are unwilling to go proselytizing among the scientifically untrained, unconvinced public. The chemical community offers no brownie points for such missionary service. Only total commitment to research within the monastery walls leads to canonization. Naturally, I would not be writing this now if I had not found some way of jumping those walls, some softening influence that allowed me to unbend so much as to write these fuzzy English sentences. I credit the Pill for this.

I

The year of my epiphany was 1969. At that time, I was an unredeemed chemist, sunk in the depths of workaholism: Professor of Chemistry at Stanford University, with one of the largest and most active research groups in the department; President of Syntex Research, then a very rapidly growing pharmaceutical company largely transplanted from Mexico to the Stanford Industrial Park shortly after my own move from Mexico to Stanford University; chairman and CEO of Zoecon Corporation, a new Syntex spin-off focusing on new, hormonal approaches to insect control; and chairman of Syva, a joint venture of Syntex and Varian, dedicated to research on organic superconductors and stable free radicals (the only kind of radical most chemists ever know). While I seemed to have found 26 hours in a day, given the steep growth curve of each enterprise, I clearly needed more. If I were to continue my professional polygamy, satisfying my first love (academic research) as well as three industrial affairs on the side, something would have to give. Syntex was then growing so rapidly (primarily because of its commercial success with oral contraceptives and with topical corticosteroids that had also been developed in Mexico in the 1950s) that everyone expected me to relinquish my other corporate interests and to concentrate on it. But I had also come to realize the depth of my belief that small is more attractive than big, so I chose the youngest company, Zoecon, for my non-academic hours. I fantasized about duplicating the Syntex experiment once more—converting a small, innovative research enterprise into an integrated operation involving research and development, manufacture and sales—and to do this in a field that could benefit society as well. On 31 December 1972, I left Syntex.

The seed for this departure was probably planted in 1969 when my first public policy article appeared in *Science*. Up until that time, I had only published factual chemical papers dealing with essentially black and white problems, amenable to experimental verification. To take on the fuzzy grayness of policy and politics was, of course, something of a departure for such a 'hard' scientist. The paper, 'Prognosis for the Development of New Chemical Birth Control Agents,' suggested that whatever the United States might do in contraceptive policy and research would have an overwhelming impact on developing countries, and that, as a matter of fundamental decency if not enlightened self-interest, we ought to take a global view of contraception. Currently, heart disease, cancer, and rheumatoid illnesses are high on the priority list of pharmaceutical companies, which focus mainly on the affluent geriatric consumers and not on the impoverished pediatric ones, whose lives could be significantly eased by improved birth control. People didn't exactly yawn when they read that article, but neither was there much excitement. The approach was too global, the recommendations too altruistic.

The following year, in 1970, I launched what I still consider my most influential contribution to public policy, 'Birth

Control after 1984,' which also appeared in *Science*. As I have already mentioned, no other paper of mine, technical or non-technical, was as widely reprinted as this one, perhaps because I examined the problem through American rather than global lenses. It starts with this observation:

> It behooves us to consider what some of the future contraceptive methods might be and especially what it might take, in terms of money and time, to convert them into reality. There are many publications on this subject, but none seems to have concerned itself with the logistic problems associated with the development of a new contraceptive agent.

I then proceeded with a detailed plan, including critical path maps (citing each separate development step and a time estimate for its completion), for the development of two fundamentally new, yet scientifically feasible, contraceptive agents: a once-a-month pill for women, and a pill for men. While 30 years later, neither invention has as yet been realized, I wasn't that far off with my first prediction for a new female pill as shown by the following quoted excerpts from a paper by E.E. Baulieu when he described the early history of the anti-progestational agent RU-486—scientifically the only truly novel development in birth control since the development of steroid oral contraceptives based on progesterone mimicks (i.e. the Pill):

> C. Djerassi in 'Birth control after 1984' (1970) defined 'as an important example of future contraceptive methodology in the female' a 'once-a-month pill with luteolytic or abortifacient properties, or both . . . [Specifically, I described it as a menses-inducing pill that women would take just prior to the expected menstrual period to initiate bleeding]. Such a compound 'may

> well turn out to be a steroid,' Djerassi said . . . but the concept of *antiprogestin* was not mentioned. Indeed, at that time, the key biological component [the progesterone receptor in the uterus—a discovery only announced later that year by Baulieu] had not yet been revealed.

I had chosen the date '1984' in the title of my 1970 article not only for its Orwellian overtones, but because I wanted to emphasize—again a point never before raised in public—that it would take an average of 14 years from initiation of laboratory research for such a new contraceptive to final FDA approval for wide public use. (Coincidentally, that was exactly the elapsed time interval between the start of the French chemical studies on RU-486 and the first articles confirming clinical efficacy.) I ended my article with a set of legal and financial incentives, without which the continued participation of the pharmaceutical industry in this field would largely disappear, and concluded with the words, 'Birth control in 1984 will not differ significantly from that of 1970.'

In late 1972, shortly before I received the National Medal of Science from President Nixon for the first synthesis of a steroid oral contraceptive, the industrialist Carl Djerassi had to sabotage the citizen-scientist Carl Djerassi. I concluded in my capacity as President of Syntex Research that it did not make commercial sense for Syntex to continue to spend money on R&D in the contraceptive field. Unfortunately, the industrialist was right: in 1969, 13 large pharmaceutical companies—nine of them American—still had meaningful research commitments in the field of birth control. By 1984, only four of them were left, only one of them American.

II

The thoughts behind these two public policy articles had convinced me that politics, rather than science, would play the dominant role in shaping the future of human birth control—a personal conclusion that not only affected my decisions as a research director in industry, but also caused a dramatic shift in my role as a university teacher—a transformation that would never have occurred without the Pill. Until then, I had taught only chemistry courses, primarily advanced ones directed at graduate students. But if politics starts to have a negative impact on a technical field, and especially on one so important to society, then it seemed to me that the most constructive action I could take was to educate the decision-makers and politicians of the future, to which Stanford University has contributed a significant number. (To cite only one example: one-third of the current US Supreme Court, including the Chief Justice, are Stanford graduates.) Precisely at that period, as it happened, an innovative new undergraduate program was established at Stanford with financial support from the Ford Foundation. The Program in Human Biology was designed to combat the increasing scientific illiteracy of our culture during a time when most public policy issues had acquired technological or scientific aspects. Much of that illiteracy can be traced to the poor quality of our high school education in mathematics and the hard sciences, which leaves our students, even at the most prestigious universities, with a persistent fear of such subjects. One way to offset this weakness is to emphasize the less quantitative areas of science, notably biology, and to do so on the most anthropocentric and thus most persuasive front: the study of *homo sapiens.* Not surprisingly, considering the composition of university faculties at that time, all the founders of Stanford's Human Biology Program were men, among them the geneticist and Nobel laureate Joshua Lederberg, the population biologist Paul Ehrlich, the pediatrician Norman Kretchmer, the neurobiologist Donald Kennedy (later to become commissioner of the FDA in Washington, and eventually president of Stanford University), the psychiatrist David Hamburg (subsequently president of the Carnegie Foundation) and several other distinguished academics. They devised a two-year core curriculum that would both enable students with minimal exposure to the physical sciences to become proficient in the biological and social sciences, and prepare them for two years of advanced courses in more specialized fields—all this to be superimposed on the regular liberal arts requirements of the university. These senior professors also served as the principal lecturers in the courses.

Within a few years, Human Biology became one of the most popular undergraduate majors at Stanford, selected by students whose goals were medicine, public health, law, environmental sciences, and politics—precisely the constituency I wanted to address on contraception and population issues. So I became, and still am, the only chemistry professor to join the Program's faculty, which otherwise bridges the 'hard' and 'softer' sciences by having distinguished representatives from biology, psychology, sociology and anthropology, as well as many medical school departments. I volunteered to offer a course for advanced undergraduates under the rubric 'Biosocial Aspects of Birth

Control'—the first 'policy' course in Human Biology and one that eventually led to a total change in my life as a classroom teacher. I chose that topic because I felt that birth control affects nearly everybody: people have used it, will use it, or are, at worst, against it.

Of my several aims, the most important was to encourage students to think seriously about public policy with real problems in mind. At a time when Stanford offered no formal undergraduate public policy courses, I felt that my professional background, bridging academia and an industry highly concerned with risk–benefit considerations, would qualify me for such teaching, especially since at that time, none of the other faculty members in the Human Biology Program had worked in industry. In proposing this course, I especially did not want to limit my audience to prospective scientists; the future legislators and formulators of public policy are unlikely to come from that guild. By using 'biosocial' in the course title, I hoped to make it plain that I was emphasizing the 'softer' and broader aspects of my subject, and thus to attract students from a wider circle. I omitted all course prerequisites other than the requirement that students be seniors, and thus competent in at least one relevant discipline: the departments I proselytized included religion, psychology, sociology, anthropology, economics, and political science; in biology and chemistry, I would find the pre-medical candidates, and for them, too, I cast my net. Never in my career as a chemistry professor had I looked for customers; now I found myself becoming a promoter. I composed a one-page broadsheet in which I outlined the purpose of my course and the manner in which it would be taught. Attached to it was a questionnaire, which every interested student was asked to complete. I was not just curious about their academic qualifications but also about their social and geographic backgrounds, especially about travel and life abroad. I had a special educational experiment in mind, for which I needed a special group: equally distributed by gender, and with adequate representation from various ethnic, social, and religious backgrounds. I limited enrollment to 40. Since over 80 students completed the questionnaire, I was able to start with a highly select and motivated group. Not surprisingly, given what I have said about the culture of academic chemistry, very few 'pure' chemistry majors applied.

I like to think that it wasn't only the subject matter that attracted the students, although 1972 was the height of the sexual revolution, and contraception a topic that either interested or antagonized almost everyone. I like to believe it was the unusual structure of the course I described in my blurb. There would be no examinations, and my formal lectures would end after two weeks. During that time, the students could pick among a variety of population groups, whose birth control practices they would then study in depth in groups of five or six. The emphasis would be on projected improvements in birth control, with each student examining the chosen population group from a particular disciplinary standpoint. A typical task force might include majors in pre-medicine, pre-law, economics, religion, anthropology, chemistry, and psychology. Each task force member would write a separate chapter of the group's report from her or his professional perspective.

The main purpose was to demonstrate that an ideal, universal birth control approach was a chimera—in retrospect, an obvious conclusion, but one that I had largely ignored during my days as a 'hard' scientist during the 1950s and 1960s. Because of the tremendous differences among cultures and individuals, what is appropriate for one may not suit the next. I wanted the students both to see that what the world needs is a kind of contraceptive supermarket, and to propose, through their own research, what items on the shelves of that supermarket might appear particularly attractive to widely differing constituencies. My first class chose to study white, American college students (typified by the majority of Stanford's affluent student population at that time); Chicanos in San Jose (a politically and economically disenfranchised group of Catholics); Puerto Ricans in Manhattan (a similar group on the East Coast); people in the lower-income strata of Mexico City; and Egyptian peasants in the Nile Delta and Indian slum dwellers in Calcutta—two third-world populations in quite different religious and political settings; and, finally, a group representing the 'women's liberation' position.

This first class in 1972 turned out to be an important educational experience for me as well as for the students. All of us worked extremely hard. After the second week, once I had completed my lectures—illustrated with many slides, and each lasting for nearly three hours—I met twice a week separately with each task force. During those sessions, I questioned each student about her or his research progress; but most important, I insisted that the students collaborate. Although in real life the important social and technical advances are all the result of interdisciplinary team efforts, we don't often incorporate that concept into our undergraduate curriculum. Instead, our entire grading and evaluation system emphasizes individual performance and competition; collaboration is cheating. Since everyone wrote a separate chapter, I had no difficulty evaluating the individual accomplishments, but these contributions had to be integrated within the entire group's report; each student had to know what every other member of the group was writing. In the student evaluations of my course, this task-force approach was voted the most original and worthwhile learning experience. Two decades later, a student from that first class, now equipped with both MD and PhD degrees, wrote what every teacher yearns to read: '"Biosocial Aspects of Birth Control" was *the* most important course I have ever taken . . . You taught me how to fish instead of simply giving me a fish to eat when I was hungry.'

The climax of the course was the task forces' presentations of their conclusions. Each group had available three hours—half for the formal presentation, the other half for questions and answers. It was during these presentations that the students really surprised me. I had given them *carte blanche* in presenting their conclusions, provided every member of the group participated in some fashion. The first task force used a mixed lecture-projection-performance format—a device that pushed some thespian button in the other groups: from then on, students used everything from skits to full-fledged dramas. Even though I taught this course only every two years during the 1970s, word spread among the students about these presentations, and subsequent classes tried to outperform their predecessors.

A task force in the mid-1970s dealing with birth control problems among black Americans gave one of the most memorable presentations. The group wrote and performed a tragic-comic drama, which effectively demonstrated several basic factors that they felt determined birth control alternatives chosen by an American black, urban population: the high teenage pregnancy rate; the non-judgmental and generally supportive attitude of black parents and grandparents in respect to teenage pregnancy; young black males' general lack of interest in effective birth control; and white social workers' relative ignorance of black family interactions. The role of the white social worker was played by a light-skinned black student, who took it for granted that the young teenage woman would have an abortion—only to find, on coming to visit the family to arrange for the procedure, the boyfriend, the girl's parents, and a grandparent all sitting in a modest living room and planning the birth of the baby. The woman who played the pregnant teenager subsequently became pregnant herself, shortly before entering medical school. I was proud when I learned later that even as a single mother she had successfully graduated as an MD.

My third class, in the fall of 1975, conducted the most ambitious projects. By that time, in recognition of the demands it imposed on its students, the course had been expanded to two quarters. I had contacted the Rockefeller Foundation to ask whether it would fund, as a one-time experiment, travel expenses for the class to conduct some exploratory research in more distant locations. Until now, support for my students' research was limited to reimbursement for telephone calls and local travel within a hundred miles of Palo Alto. Students who had chosen more distant population groups had to depend on what they could cull from phone calls, the library, or memories of past travel. The Rockefeller Foundation, being especially devoted to supporting research in developing countries, agreed to fund this educational experiment. As a result, I was able to organize the largest class of all—with ten task forces—and to offer each group the opportunity to send at least two, and sometimes all, members to sites at any distance. The three most geographically ambitious projects involved populations in Kenya, Java, and rural Mexico, but the American-oriented projects were also interesting. For instance, the Chicano task force, consisting of four students named Martinez, Ramos, Renteria, and Rios, decided to conduct a comparative study of Chicano communities in Denver, El Paso, and Los Angeles. Several members of another group, which chose native Americans, spent a couple of weeks around Christmas in New Mexico with an Indian tribe. (Taos also has great skiing, which sent one of the task force members home on crutches.) A third group picked a rural setting in the South—Cherokee County, North Carolina, the home of one student—and provided a generally unfamiliar view of the educational and public health restrictions operating there.

Two particularly interesting choices had no geographical, but rather a functional definition: one dealt with the birth control problems of carriers of genetic diseases; the other, with those of the developmentally disabled. Reading all of these reports was a monstrous task, and to address it, I secluded myself one drizzly winter weekend at my ranch home. Soon wearying, on the spur of the moment I picked up my umbrella and took to my outdoor hot tub, where I floated naked without dropping a single page into the water. Of course, the steam was not without effect, but I never confessed to the students why the pages of their reports came back slightly curled.

The oral presentations of these groups were on the whole impressive—luckily, for so was the composition of the invited audience. The medical director of the Rockefeller Foundation flew out from New York for several presentations, and the chairperson of the California State Assembly's subcommittee on health came to those on genetic disorders and the developmentally disabled—spurred by then-pending hearings on those topics in Sacramento. The report of the Indonesia task force, which dealt with market and social incentives for contraceptives in Java, got one of the students a job with the Agency for International Development in Washington; on the basis of his field report, the World Health Organization hired a member of the Kenya group for a summer internship in Geneva before he entered medical school. Sharon Rockefeller, the wife of the current senator from West Virginia and a trustee of Stanford University, attended the presentation of the Kenya task force; when the student performers passed around fried termites, a Kenyan delicacy, I seem to recall that Mrs Rockefeller was one of the few to take up the offer.

When I started teaching the birth control class, the distribution of the students by gender was approximately equal. By the late 1970s, fewer and fewer men were choosing to enroll; and by the early 1980s, at most 20 per cent of the students were men. This gender tilt was even more pronounced in an offshoot course I first sponsored in 1983 and have since taught every two or three years. I had observed that the one task-force topic always selected by each class dealt with what I called at that time 'the women's liberation position.' Aside from its obvious timeliness, there was another reason it was such a popular choice. In my lectures, I always drew attention to the absence of new male methods of contraception and to the possible reasons for such an omission. Invariably I asked my women students what might be the results if they had the power to make financial and manpower allocations. Asked to choose relevant as well as *realistic* research targets, most of the time, they found hard-nosed scientific reality led them toward their own reproductive organs. As the years passed, and especially when Stanford introduced a special Feminist Studies Program, I decided to create a new course to address these questions under the title 'Feminist Perspectives on Birth Control.' I have now given it eight times, once at Bard College in New York. With a solitary exception, all the students in those classes were women.

But despite the opportunities I have found in working with women on these questions, I still wonder: why did men stop signing up for the course? Have men suddenly stopped believing in birth control? Has the present generation's preoccupation with material goods and professional advancement made the men relegate birth control to a low priority? I believe that the real reason is something else: since I started teaching this course, there has been a generational shift. In the 1970s the Pill was new; decisions about its use were unprecedented, forcing students of that era to look wherever they could for guidance. Since the 1980s, however, most of the students in college are children of the Pill generation. The Pill has made many important social contributions, not the least of them that birth control has become an accepted topic of dinner-table conversation.

But, concomitantly, it has also created a social atmosphere in which it has become easy to forget, and thus one more responsibility that has fallen on the shoulders of women. Many women, of course, accepted this responsibility eagerly as an important sign of emancipation and freedom from male dominance, but one of the consequences of that achievement has been a collective shrug of male shoulders, an outcome I deeply regret.

I find it both disheartening and amusing that it was the women in my class who played the biggest role in making condoms available at Stanford University. Knowing that approximately 40 per cent of all condom purchasers are now women, I thought it only appropriate to encourage some of the students in 'Feminist Perspectives' to focus on that form of contraception. Among the papers women in my class wrote were critical evaluations of condom advertising and promotion. For instance, instead of a phallocentric terminology like 'Sheik' or 'Ramses' or 'Trojan', why not call a brand of condoms 'Cleopatra'? And instead of blue and green and orange, one of my feminist warriors in class asked, why not color the 'Cleopatras' gold (a color that has since become popular)?

On a perhaps more practical note, in 1980, two women task-force members examined what it would take to introduce condom dispensers at Stanford University. Their reward was (among others) a first-class lesson in the function (or lack thereof) of an academic bureaucracy. The dean of student affairs sent them to the acting deputy vice president for administrative services and facilities, who suggested they first see one of the university's legal counsels, and then (for the inescapable reason that they were thinking of putting the dispensers in gyms) the athletic director. When the students suggested the library's toilets as a suitable site the horrified librarian's response was, 'Just imagine all the high school students who would come to the library to get condoms!' At that point, not even the university's ombudsperson was able to help.

I could hardly think of a better use for condoms (or a better way of getting high school students into a library), even at the price of finding a used condom or two among the stacks, but no one was asking for *my* approval. Not until 1987 did a feminist Gang of Five, led by my students Shirley Wang and Jennifer Yu, succeed in wearing down Stanford's administration and getting the first condom dispensers into some of the toilets. Were it not for the fact that I practice what I preach by having had a vasectomy many years ago, I would have been one of the first customers. Even so, I keep a huge collection of condoms for class demonstration purposes, including such curiosities as a box from Kenya. I am probably one of the few persons who could claim the cost of condoms as an income tax deduction for my professional activities as a teacher—one of the many benefits I would never have dreamed of enjoying, had it not been for the Pill.

III

The first 20 years of this softening of my scientific persona transformed not only my teaching, but also increasingly my writing. In addition to the usual research articles, co-authored with my graduate students and postdocs, I was sole author of over a dozen policy papers directed to a general audience that culminated in the earlier mentioned book, *The Politics of*

Contraception, first published in 1979. In propelling me toward a general audience, the Pill was opening a door. Before then, it may have been unlocked or even slightly ajar without my realizing it, but it has now swung so wide that on occasion I am still startled at where it has led me. Although the motives behind my further transition from a softened chemist to an author of fiction are complex, as I have described, I feel justified in crediting even that in part to the Pill. My last two novels, *Menachem's Seed* and *NO*, and my first play, *An Immaculate Misconception*, all deal with human reproduction. I am convinced that I would never have focused on that subject had it not been for the intellectual push into that field provided by the Pill.

Fiction writing has led me in turn to try another new pedagogic experiment. I am now the first chemist at Stanford University to offer a course in our medical school under the auspices of the Center for Biomedical Ethics. Given the recent proliferation nationwide of new courses on biomedical ethics, what is different about 'Medicine 256', as it is listed in the catalog?

The impetus for this course, 'Ethical Discourse through Science-in-fiction,' was the unanticipated success that my first science-in-fiction novel *Cantor's Dilemma* enjoyed as an academic textbook. The paperback is now in its thirteenth print run primarily as a result of adoptions by American colleges and universities. In the majority of cases, the novel is used in courses dealing with ethics in research, because the novel's plot describes a dubious misstep in research. Graduate schools of business administration have long learned the advantage of employing 'case histories' in their curricula, which makes the adoption of *Cantor's Dilemma* easily understandable. But ethical or behavioral deviance in research generally involves individuals rather than impersonal entities such as corporations, which means that scientific ethical case histories quickly run into concerns about violation of privacy as well as into manifestations of the whistleblower syndrome. I wondered whether student-generated 'science-in-fiction,' in which all aspects of scientific behavior and of scientific facts are described accurately and plausibly yet disguised in the cloak of fiction, could be used to illustrate ethical dilemmas that frequently are not raised for reasons of discretion, embarrassment, or fear of retribution.

'Medicine 256' was restricted to graduate or postdoctoral students, because I wanted stories based on experience, or at least on events that the writer knew enough about to fictionalize with authority. The first time I offered that course, 14 graduate students and postdoctoral professionals from 12 different departments enrolled to compose short stories dealing with ethical issues in science or medicine. The stories had to be handed in on the first day of class (meaning that they were written during the preceding quarter break) and were then distributed to all class participants without authorial identification —thus permitting unrestricted debate. The balance of the course dealt with in-depth examinations of the ethical or behavioral problems raised by these stories—discussions that frequently bordered on fireworks. In all my 40 years of teaching at Stanford University, this was the most exciting classroom I experienced. Aside from offering the students a veil of anonymity that removed most obvious restraints, the course had allowed them to pick the topics that concerned them most rather than being restricted to ethical problems chosen by the instructor.

(Since then, I have been tempted to organize such a course on a transnational basis to see what problems similar students from say Italy, Germany, and Japan would select for their stories compared with the concerns created from work in the pressure cooker of an American elite university.)

Participants were judged on the subtlety of the ethical issues raised, rather than on the literary quality of the product, but both my wife—a professor of literature—and I were amazed by the quality of some of the writing. But on further reflection, I realized that I should not have been surprised because of the self-selection that this course had inadvertently demanded. Many doctoral thesis supervisors were not too sympathetic to their students taking what they considered a 'Mickey mouse' course: instead of spending time on literary fantasies, why not spend those precious hours in the laboratory? Such attitude is by no means unique to professors. It is shared by many students in the hard sciences such as chemistry, where 'ethics' courses—if taught at all—are not taken very seriously, the tacit assumption being that 'hoaxing, forging, trimming and cooking' in research is a pathology encountered mostly among biologists or clinicians. I have now taught 'Medicine 256' three times, with students from well over a dozen departments, but during those years only two chemists, both female, dared to cannibalize a small portion of their 80-hour work week for such a 'soft' exercise. To be tempted by such a course, the students had to be not only strongly self-motivated to do some non-scientific writing, but also deeply concerned or even wounded by some behavioral aberrations in academe. One such story—among the best from that first class—appeared subsequently in a medical publication, anonymously (at the request of the female author). Like two other stories, it dealt with human reproduction, though not the Pill. In this one, a young physician finds herself in the painful role of having to advise a semi-literate, pregnant Albanian refugee to undergo an abortion, while she herself is facing the heavy decision whether to abort her own first pregnancy. The story was autobiographical. Several students had tears in their eyes when it was discussed in class.

In addition to creating a forum for open disclosure and debate, my course also addressed the question of how scientists might communicate better with their colleagues and the general public. This discussion led to an extraordinary exercise. I had the class attempt a group composition modeled after the Japanese Renga (linked verse in which stanzas are composed by two or more poets in alternating sequence, often as a form of competition) for a short story dealing with some scientific ethical dilemma. Each paragraph was to be written by a different student without knowing the identity of any predecessor author, each being allowed two days to compose her or his paragraph. Once the 14-paragraph 'Science Renga' was completed, each student was asked to add a fifteenth paragraph, thus generating 14 new endings. After distributing all variants to the class, the 'winner' was selected by closed ballot. Though bearing the names of all authors—a feature common among scientific papers but virtually unheard in literary publication—none knew who had contributed what segment.

Renga bears an interesting resemblance to the process of scientific co-authorship, which also has its collegial and competitive aspects (indeed I used the idea in my second novel, *The*

Bourbaki Gambit). But the Renga experiment of the class was a 'purer' collaboration, since each author was associated with the entire enterprise though no identifiable individual component. I decided to see whether a scientific journal would have the guts to publish this story by starting on the very top, with *Nature*—the science equivalent of submitting a first short story to the *New Yorker*. But the *Nature* editor bit within a week (itself virtually unprecedented) and 'A Science Renga' under the names of 15 authors appeared in the 11 June 1998 issue of the journal. It was the first piece of fiction that *Nature* had ever published—at least knowingly so—since its founding in 1869. So unusual was this event that a major French newspaper, *Libération*, featured it on a full page. It may also be the first short story in literary history that bears the name of 15 authors. But why 15, when there were only 14 students?

Though dangerously tainted by overtones of a shaggy dog story, the answer is relevant to the subject matter of 'Medicine 256', since 'senior' authorship is so often the source of ethical conflicts in science. Whose name should come first? Some students suggested the device of 'honorary' authorship, so common in science yet still not sufficiently condemned; in other words to compromise with 'Djerassi *et al.*' 'It was your idea,' they said. 'You organized it and even got it accepted in *Nature*. Put your name in and put it first.' Of course I rejected that alternative since adding one's name to papers where one did none of the actual work was one topic the entire course had meant to address. Alphabetical order was the next obvious alternative, a common enough approach that is also fraught with complications. Here is one such case, taken from *Cantor's Dilemma*.

> 'Most people in the field—including Celestine—would consider me the senior author. It's not necessarily the first name in a list of authors, although some senior researchers feel very strongly that their names must always appear first. Others always use an alphabetical order . . .'
>
> 'Well, it's not true in our lab,' Stafford mumbled, 'it's always alphabetical.' This was the only serious bone of contention in Cantor's group. Lab gossip had it that no Allens or Browns had ever worked with Cantor. There had been an exchange fellow from Prague, named Czerny, but that was the closest alphabetical proximity to 'Cantor' that anyone remembered until Doug Catfield had arrived last year.'

Alphabetical precedence among the fourteen authors of 'A Science Renga' would have meant that Dina L.G. Borzekowski, a postdoctoral fellow from the Stanford Center for Research in Disease Prevention, would appear as senior author in the *Nature* index. But was that fair, when all deserved equal credit? Ordinary mortals outside the scientific community are often astounded by our preoccupation with names on papers and the complicated solutions we sometimes devise, especially since many professors—especially 'honorary' authors—now place their names last. But what about the front of the queue? Here is another relevant excerpt from one of my novels, this time the last volume, entitled '*NO*.'

> 'What we were talking about, before you came,' Celestine turned to Paula, 'was the subtlety of how to apportion credit among all the authors. These days, four or more authors is par for the course in any competitive field of chemistry or biology. Having settled who is last, the question now is who comes first.'

'Really?' said Paula. 'I'd think you'd pick the person who has done most of the work.'

'You think that's easy? That's exactly what we've been discussing. Recently, John Scott from Portland published a real first in *Science.* He had five co-workers, all women—a real harem—but what made it a first was that the first two names listed in the article were marked with an asterisk. Can you guess what the footnote said? "These authors contributed equally to this manuscript." '

'Brilliant,' exclaimed Paula.

'You see?' laughed Marletta.

'Brilliant?' Celestine snorted. 'Suppose the first asterisked name had been Smith and the second Price. I would have gone to Scott to point out that in any citation, that article would be referred to either as "Smith et al" or "Scott et al." To me, "et al" does not mean "equal." '

'So what would you have had Scott do?'

'Ah,' grinned Celestine. 'As a first try, I'd have separated the names Smith and Price by an equal sign rather than a comma. But since no editor would allow that, I'd have told him to do it alphabetically.'

'You mean Cantor's system? Why should Smith agree to that when your name starts with a P?'

'Fair enough. That was also Michael's point. So I asked why not toss a coin? And you know what the fair-minded Professor Michael Marletta said?' Celestine poked him lightly with her index finger. 'Why don't you tell Paula.'

'In my lab, I decide such issues, not the drop of a coin.'

I felt that none of these conventions would work with our Science Renga. When the paper appeared in print, my unilaterally chosen first author was one 'Alfred N. Alston Jr,' whose name was followed by an asterisk that did not indicate a department address but rather the fact that he was deceased.

Nature never caught the discrepancy—14 students, but 15 authors—but an interviewer on a radio show did, and asked me to explain. 'It's an anagram,' I admitted, and then challenged the listeners of the program to come up with the answer by e-mail. The first correct respondent was promised a signed copy of *NO*. I had barely returned home to find a correct answer: 'Leland Stanford Jr,' the person after whom our university is named, and now the author under whom 'A Science Renga' will be found in perpetuity in the annual index to volume 393 of *Nature*.

The student authors were so jubilant to find their names in *Nature*, an addendum to their professional biographies that many of their professors could not boast of, that none objected to my unilateral decision about senior authorship.

I am almost finished with my drawn-out record of this latest pedagogic experiment I credit indirectly to the Pill—but not quite. One of the participants in my course, a talented poet, E. Weber Hoen, decided to compose an abstract of the original 'A Science Renga' (since abstracts are *de rigeur* in every published scientific article). But he did so in the form of a Shakespearean sonnet, where each of the 14 lines corresponded to one of the paragraphs of the Renga! The 'old goat' of the title is the professor in the short story, who was terrified of being scooped by his younger disciple. I quote the last six lines of Hoen's sonnet:

Old Goat

It is height you desire, and with that, truth,
to shake your beard on an eternal view,
as if from there you might behold your youth.
The rain, though, has you blind. Below, like you,
the young conspire in fear against their king.
Goat, you are old. You have not learned a thing.

When I first read the sonnet, I realized that my age clearly qualifies me as an 'old goat.' But I am luckier than Hoen's old goat. The Pill has taught me a thing or two.

Chapter 10

The Pill and Paul Klee

On 30 July 1985, a *San Francisco Chronicle* front-page article bore the headline, '6 Stolen S.F. Museum Drawings Found.' At that time, the San Francisco Museum of Modern Art, commonly known as SFMOMA, had not yet moved to its elegant new quarters south of Market Street. It still occupied the two top floors of the War Memorial Building, next door to the San Francisco Opera, its main entrance facing golden-domed City Hall across Van Ness Avenue. The War Memorial's ground floor entrance hall is a rather grand space that is occasionally rented out for anniversary celebrations, weddings, and other private functions. On the less fashionable side of the street, SFMOMA faces some apartment buildings, one of them the abode of a 19-year-old psychology student attending San Francisco State University. When an art-student friend from out of town paid him a visit, his host suggested they crash a private party then in progress in the War Memorial's foyer.

'You know, being a student with no money, it's an easy way to get free food and drink,' the newspaper quoted him. 'So me and my friend went over.' According to the thief, the reception was saving the champagne and food until later, so they started to get drunk on vodka and grapefruit juice. Some people get sleepy when drunk, others belligerent, but some art lovers evidently turn curious. 'We wanted to see what the museum looked like in a different light,' the instigator confessed, so they climbed the rear steps to the top floor and pulled on the door until 'it just popped open.' (Even now, 15 years later, I get weak-kneed when I visualize that scene.) 'I hadn't seen Paul Klee for a long time, so we went over to where his paintings are,' said the discriminating burglar, 'but it was dark, pitch dark, so we took four off the wall to a hallway so that we could look at them.' On the way out, they passed some Picassos. 'My God,' he said, 'Picassos! Let's take two of them.'

Unbeknownst to the drunks, they had set off a silent alarm, but by the time the private security personnel had come to check, the thieves were gone with six drawings under their coats. The keystone cops working for the firm, who, I trust, has since fired them, found nothing amiss. Around 3:00 a.m., the two Klee kleptomaniacs had started to sober up and panicked. Deciding to return the stolen drawings, they made the return trip across the street with the four Klees and two Picassos under their coats, but found the front doors hermetically locked.

Upon reading in the morning's newspaper about the value of their loot and the magnitude of the sentence they were facing, the younger of the now cold-sober pair left the art works in a cardboard box on the third level of a downtown parking garage and called the police.

I

The story was rather funny in a slapstick sort of way—except to the embarrassed museum officials and the pained owner of the purloined art. That owner is me. And what has the Pill to do with all this? The answer is simple: the money that made possible the purchase of those Klee drawings (and well over one hundred others that have been shown at numerous Klee exhibitions at SFMOMA) all came from the Pill.

Not directly, of course. Although I never received royalties from the Pill itself, my tenure at Syntex was rewarded—as a great many start-ups prefer to compensate their ambitious young talent—with stock options. In fact, so much did I believe in the work I was doing that I also purchased Syntex shares on the open market. In my case, the promise of future reward was realized, and handsomely: the value of those shares and options appreciated greatly when the market came to recognize the commercial potential of the Pill. During the 1960s, after Syntex had metamorphosed into a public company traded on the American Stock Exchange, its stock turned into one of the hottest commodities on Wall Street. By the end of that decade, against all my expectations, I looked around and found that chemistry had made me rather affluent.

But wealth was never for me an end in itself. As I grew more knowledgeable about art, I became the typical collector: obsessed, and increasingly selective. For a long time, I focused on works by artists who were both painters and sculptors: Degas, Giacometti, Marini, Picasso and others of that class. Twenty years later, I sold them all, in part to practice another form of art collection, the support of living artists through the creation of a resident artists program at my ranch in the hills west of Palo Alto. After all, collecting the works of dead artists does nothing for the person up in Parnassus, whereas living artists need time and resources for creativity to flourish.

I sold a great deal of art to support that project, but I always made one exception. Almost as soon as I could afford it—in the mid-sixties, when prices were also much lower than today—I began collecting works by Paul Klee. He has been dead since 1940, yet I still collect Klees, and have not sold any. I did not purchase Klee's works for financial returns. Rather, I was fascinated by his versatility, the range of media he chose, and his deep intellectuality, the latter manifesting itself not only on his canvases, but also in his writing (pedagogic, autobiographical, and poetical) and even his musical appreciation. In over 9000 works, he used every conceivable surface from paper and canvas to glass, burlap, gypsum, and even ceramic. His stylistic experimentation did not just change with time: Klee frequently worked simultaneously on a dozen pieces, which might vary enormously. I am certain that the thieves, even if cold sober, would never have guessed that the water color *Mazzaró*, one of my all-time favorites, and the oil-transfer drawing *Oriental Girl*, both of which they stole in July 1985, were all completed within weeks of each other in 1924.

II

In my college days, I had seen reproductions—on postcards, calendars, posters, catalogues, and art books—of Paul Klee's most famous works, like the *Twittering Machine* or *Ad Parnassum*.

Subsequently I saw many of his oils, drawings, and watercolors in museums in Europe as well as in America. In the mid-1960s, I went to my first Klee show in a gallery in London, where all of the works were for sale. I kept returning to two magnificent watercolors from his Bauhaus years in the 1920s—rather large ones for an artist who usually worked on a small scale. 'Should I? Can I?' I asked myself, and realized for the first time that I actually might be able to afford one of them. Finally, I approached one of the gallery employees and asked about the price. 'The 1925 *Horse and Man*?' he asked, looking me up and down. 'Sixteen,' he finally said.

'Sixteen what?' I wanted to ask, but didn't. I knew it could not be 1600, and was unlikely to be 160,000, so it had to be 16,000. But sixteen thousand what? Dollars, pounds, or even guineas? 'And the other one, the 1927 *Heldenmutter*?' I asked hesitantly.

'Eighteen.'

'Hm,' I replied and went back to look at the pictures. A few minutes later, the man appeared by my side. 'Which one do you prefer?' he asked in a slightly warmer tone.

'I can't make up my mind. Both are superb.'

'Buy them both,' he said matter-of-factly, 'and maybe we can arrange a better price.'

Bargaining, whether in a market-place in Mexico or a bazaar in Cairo, always makes me uncomfortable; but this time I haggled like a rug merchant. Every retreat of mine, every inspection and reinspection of first one and then the other Klee caused the price to drop. They were not big reductions, but given the overall sums—far above anything I had ever spent before on art—they were not insignificant. Finally, I said, 'I'll have to think about it.' A couple of days later, I was the owner of two Klees. Even now, after decades spent collecting Klee in all his various media, these two are still among the *crème de la crème.*

My purchase of these first two watercolors was exhilarating for me, but at the time I had no idea if it was merely a fling, or the beginning of something more serious. All doubts on that point were resolved with the acquisition of my third. I saw it on the walls of the Guggenheim Museum; it appeared with the notation 'Collection Galerie Rosengart.' It did not take me long to secure the address of that gallery in Lucerne, or to consummate by mail the purchase of that small gem—a description I learned eventually that Klee himself agreed with. Galerie Rosengart was owned by a father-daughter pair, Siegfried and Angela Rosengart, among the most important dealers and collectors of Klee's works. Eventually I got to know them well, and made frequent pilgrimages to their gallery. A couple of years after I bought the watercolor off the Guggenheim walls, Rosengart *père* told me the significance of the small notation 'S Cl', which Klee had marked in pencil in the lower left-hand corner of that watercolor. An abbreviation for *Sonderclasse*, 'special class,' it denoted his own favorites among over nine thousand works. I later learned that Klee had marked earlier works 'S Kl,' until someone told him that spelling *Classe* with a C had, so to speak, more class.

A few years later, I expanded my collection to include Klee's graphics, of which there exist only about one hundred. The early ones, done between 1901 to 1905, are Klee's first truly original creations and also among his rarest. He called them 'inventions,' and titled them ingenuously, such as the famous 1903 'sour' print, 'Two men meet, each believing the other of higher rank.' Their inherent sarcasm; their bizarre portrayals of the human figure; even their frequent perversity—these are some of the reasons why I continue to search for additional works from that period. In the middle 1970s, there was a major auction of Klee graphics in Bern, where Klee spent his last years, where his son Felix lived until his own death, and where, within the walls of the *Kunstmuseum*, the Klee Foundation is situated. By that time I knew most of the important Klee dealers as a customer and, in a couple of instances, as a friend. I had studied the auction catalogue and carefully examined the lots prior to the start of the auction. I had a mental art budget for the year, which promised—with luck—to suffice for the purchase of two of the early graphics. At an auction, however, one never knows. All it takes is for two people to become enamored with the same lot, and the price heads for the stratosphere.

I informed two of the dealers I saw in the audience that I was going for these two etchings; although I could not ask them openly not to bid in competition with me, I was reasonably sure that they would not drive the price up once they knew of my plans. But then my heart sank. In the distance, I saw Heinz Berggruen (then an important art dealer in Paris), who, I knew, not only had a magnificent personal Klee collection but bid actively at auctions. I approached him and, after exchanging the usual pleasantries, mentioned the two lot numbers I planned to bid on. 'Good choices,' he acknowledged. 'I was thinking of buying these for the gallery.' There was no question about Berggruen's ability to outbid me any time, but he must have read the disappointment in my face, for he offered to bid for me, saying he would just charge me a commission for any successful bids. I accepted immediately, thinking that such a commission was a bargain insurance premium in return for not having the art dealer as a competitor. I left the auction in shock. Under Berggruen's whispered persuasive urging I had bought not two, but seven Klees, blowing in minutes my art budget for five years. Yet I have never regretted following his advice. Several of the prints I bought have never again appeared at sales; and although I bought the prints purely for delight, my pleasure is not tainted by the fact that they have increased greatly in value. At the auction, Berggruen pointed out to me a man sitting behind us, whom we had outbid, as Felix Klee, the painter's son. Two years later, I visited him in his flat in Bern and saw his extraordinary collection, which included puppets his father had made for him—a genre of Klee until then unknown to me. He also showed me his mother's guest book. The first entry was Wassily Kandinsky, who did not just inscribe the book but drew a colored picture on that page. Not to be outdone, many of the other guests—Lyonel Feininger, George Grosz, and others I do not remember—did likewise. It is one of the most intimate and exquisite documents of European art of the 1920s and 1930s.

Both auctioneers and collectors can tell all kinds of auction stories—amusing, bizarre, dramatic, occasionally even tragic. I have written an entire short story around one episode, a story entitled *The Futurist* that eventually became the title of my first published collection of short stories. But I have never described publicly the story of a Klee purchase I made at a Sotheby's auction in London while stark naked. For reasons of convenience and anonymity (even at the risk of interrupted sleep) I prefer to bid at auctions over the telephone. Given the eight-hour time difference between San Francisco and London, I have on occasion been awakened around 3:00 a.m. to bid for an item in a sale conducted at a civilized morning hour in London. On this particular occasion, Sotheby's had notified me that they expected the Klee I was after to come up for bidding around 3:30 in the afternoon, London time. This is about the time I start each morning with 30 minutes of exercise on my cross-country skiing machine—one of the few forms of strenuous exercise I can do with my stiff left knee (fused from a skiing accident)—and I always do so naked before showering. It was barely 7:00 a.m. when, still puffing and sweating, I was called away from my exercise machine by the insistent ringing of the telephone. The Oxbridge accent on the phone apologized for calling half an hour early, but the auction had progressed more rapidly than anticipated. I was panting heavily, which the man surely took for excitement. The moment bidding started on 'my' Klee, I impatiently panted 'yes . . . yes . . . yes.' I soon found myself standing, sweating, shivering, stark naked, engaged in a *mano a mano* struggle with an invisible counterbidder across the Atlantic. Under the circumstances, it is perhaps not surprising that the bidding progressed at an exceptional speed to an unanticipated height: I had somehow lost my usual knack for waiting until the last moment before raising my bid. Still drenched in sweat, but now thoroughly chilled, I slammed down the phone the moment I had heard the final knock of the auctioneer's hammer to turn on the hot water in the shower.

How deeply Klee is woven into the fabric of my life is suggested by the truly unique set of coincidences that accompanied my acquisition of his *Schläfriger Arlecchino* (Sleepy Harlequin), a delightful gouache of 1933. The work did not sell at a 1995 autumn auction at Christie's in New York, whereupon I made a serious, though lower, offer to the owner, which was accepted after some minor haggling. I was pleased by my newest acquisition and promptly hung it by the entrance to my study; every time I headed to my computer I looked affectionately at my sleepy harlequin. A few weeks later, I flew to New York for a tête-à-tête with two editors from Penguin, whom I had never met before, about the forthcoming paperback publication of my second novel, *The Bourbaki Gambit.* The telephone message on my answering machine identified the restaurant for our lunch meeting as 'Arlecchino' in lower Manhattan. A good omen, I thought, as I bade goodbye to my own Arlecchino on the way to the airport.

I was a few minutes early for our 12:30 appointment at what I was convinced would be my 'lucky' restaurant. When I arrived, I found the place deserted. This could have been understandable in Madrid, where lunch starts at 3:00 p.m., but surely not in New York. 'I'm afraid we're closed today,' the sole employee of the establishment announced, 'the chef is sick.' 'Impossible,' I countered, with all the arrogance of a proud though hardly best-selling author, 'I have an appointment with my editors from Penguin.' Somehow, I thought that the plural would impress him, but my conceit was promptly punctured by the announcement that the senior editor had just phoned, claiming to be sick, and that the junior member of the team would be late. When the latter arrived, the Arlecchino employee took pity and suggested a supposedly good Italian restaurant whose name

I have long forgotten. My Penguin host and I started for the new location, but as soon as we turned the corner, I saw a sign for another establishment, 'Rocco's.' 'We've *got* to go there,' I exclaimed. As soon as we had sat down, I asked the *maître d'* to look at what I was fishing out of my briefcase. It was the bound galleys of my next novel, *Marx, deceased*, that I had brought along to show to the Penguin editor. Puzzled, the *maître d'* looked at the cover of the elegant Magritte painting showing a double-faced man. 'Why are you showing me that?' he asked. 'It's a novel,' I said. 'I wrote it. Just read the first few sentences.'

One look at the opening sentence changed his expression:

> 'Rocco's.' The voice on the telephone was brusque, the 'r' rolling like a Ferrari in first gear.
>
> 'I'm calling about a dinner reservation,' he said. 'For two. Next Thursday, 7 o'clock.'
>
> 'Name?'
>
> 'Marx.'
>
> 'Spell it.'
>
> God, not again, he thought. It's just a four-letter word.

'Amazing,' Rocco's *maître d'* said. 'Now, are you ready to order?' I waved away the proffered menus. 'Just tell us what your special is for today's lunch.' 'Gnocchi,' the man replied. It was my turn to look flabbergasted. 'I don't believe it,' I said as I turned to page seven of my novel. 'Just read this: "The gnocchi on Marx's fork had nearly reached his mouth, but at the last moment he put it down."' By now, I was ready to believe in predestination, kismet or just plain karma. But alas, not even the charm of Klee-Arlecchino-Rocco-gnocchi was able to transcend the utterly poor judgment Penguin displayed in turning down the paperback rights to my most literate novel. The only balm to my hurt ego was provided by some reviews of the hardback edition, such as the *Washington Post* reviewer's impeccably perceptive 'A classy, easy-reading page turner, light of heart and bright of mind . . . a literary novel to be reckoned with.' Even my favorite European newspaper, *The Herald Tribune*, reprinted the review, without, however, any impact on the stony, penny-pinching Penguin operative in charge of paperback rejections.

III

But why collect art in the first place? At a very basic level, people collect—art, cars, dolls, matchbook covers, even other people—to fill a void. It has taken me a long time to realize that emotionally, much of my life has been occupied by such Sisyphean collecting. While still an impecunious student, I hated empty walls. I considered art, or shelves full of books, the only acceptable embellishment, because art and literature allow a degree of personal subtlety—a surprisingly accurate definition of the one who chose them—unique in this world. To this day, when I visit a new home, I judge my hosts by their walls. Bare walls, like bare minds, turn me off.

While all art collectors acquire art, not all acquirers of art are collectors; a collector's relationship with Art—with all the connotations that capital letter implies—is complex. Collecting requires a combination of knowledge and esthetic judgment that are not necessary for the simple acquisition of a work. Collectors are patrons of living artists if the latter are still living. But when the artist of the collected work is dead, and especially when the artist has been fetishized—become a name, an object of desire, a token of exchange, a whisper of immortality—then we are entering the territory of the Arts, a place where the void that the collector seeks to fill, and the ultimate redemption of that void co-exist in some ineffable relationship. I wish I could communicate in plain words just what I feel that relationship to be—but if I could, I suppose there would be no need for art, or for collectors. I can only suggest it here, by telling that I have felt it at widely disparate moments of my life. I felt it when I found myself 'promising' (a word that promises a great many things) my Klee collection to the San Francisco Museum of Modern Art, and felt that I had become a kind of patron of the Arts.

Admittedly, my own definition of art collecting is somewhat narrow. I am not referring to the purchase of occasional pieces of art, or to collecting as investment. I am thinking of the person who concentrates on a specific artist, a specific art movement, or imposes on himself some other more-or-less arbitrary criterion; who renders an intellectual judgment and to that extent places a personal signature on the collection. Assembling five Picassos at random is very different from deliberately selecting five Picassos to make a specific esthetic, intellectual or personal point about the artist. Collecting the works of dead artists becomes a form of patronage only when it serves the public benefit. In many respects, the serious collector of a dead artist's work also becomes that artist's curator. If one takes this role seriously, then such collecting can become an exciting creative process: one presents a special view of the artist by selecting specific aspects of that artist's work. At the same time one can do something for art appreciation and enjoyment by a wider public—as well as influence developing artists—by sharing that collection. I have always deemed it unfitting when a person with the means of amassing a substantial collection—especially of a single artist's *oeuvre*—retains it for purely private gratification. Some years ago, I decided that the bulk of my Klees should go to a public museum. That conviction, in turn, has provided an additional—now, didactic—motivation to my on-going attempts to fill certain chronological or artistic gaps in my collection.

None of Klee's works is large—usually measured in inches, not feet—yet each is full of complexity, as befits the ultimate master of the *petit format*. This is incidentally another aspect of Klee's appeal for the obsessive collector: I can always find wall space for one more jewel. In fact, the size of Klee's works requires intimate space, and intimate space immediately leads to intimacy of another kind: close inspection of the work, attention to detail, and—the ultimate punctuation mark—the gasp of delight. It was the San Francisco Museum of Modern Art's promise to provide such permanent exhibition space which led me to offer the bulk of my collection in the form of a promised gift—my token of appreciation to my adopted home city. 'Home city' has special resonances to an immigrant and one-time refugee—a feeling hardly unfamiliar to Paul Klee, who himself was never totally accepted by the two countries where he spent his life.

But my deepest experience with the relationship between art and its collection, and my practical concern with patronage of artists rather than Art originated elsewhere. While that activity also requires money—another connection to the Pill—the pivotal cause for that type of patronage was the most tragic event of my life.

IV

In 1965, when my daughter Pamela was 15 and my son Dale 12, we had agreed to spend a goodly portion of my Syntex-generated wealth on some spectacularly beautiful land in the Santa Cruz Mountains, before developers discovered and ruined it. There were redwood forests, deep canyons, sweeping views of the Pacific, deer, coyote, bobcats—even the occasional mountain lion; it was only a few miles from Stanford University and an easy commute from a metropolitan area inhabited by several million people. In less than an hour, one could drive from the San Francisco Opera House into the magic solitude we had named SMIP. At first the letters stood for 'Syntex Made It Possible'; later they came to signify four more consequential words. The first 95 acres we purchased were deep in the forest; over the next few years, we acquired additional parcels that extended from the redwoods, through clumps of madrone and oak trees, to the undulating meadows pushing toward the coast—a truly feminine landscape of breasts, thighs, bellies and buttocks, the grassy skin tanned golden-brown in the summer and colored lusciously green in the winter and spring. By 1970, SMIP had become a 1200-acre spread—two-thirds of it held in the names of my children—on either side of Bear Gulch Road, a winding county road ending at the property of our neighbor, the rock musician Neil Young. On the eastern half of the property, which rises to over 2000 feet and drops to 800 feet within our own confines, we erected a 12-sided barn and a ranch manager's residence, centered on open grazing land, which became the site of a purebred, polled shorthorn cattle operation. Twelve-sided barns are rare, but the impetus here was chemical. Because the barn site was visible from various hilltops, I wanted the roof to be a hexagon, the organic chemist's favorite six-membered ring so prominent among the steroids. But since the barn's sides were mostly open to allow cattle to enter and leave *ad libitum*, the architect reasoned that Pacific winter storms might rip off the roof if it were only supported by six columns. Consequently, he doubled their number to create a dodecagon—a cyclododecane ring system that a chemist finds much more difficult to synthesize. Around that difficult ring my family began to crystallize a cluster of three homes that was to be the scene of events that changed my life forever.

On the edge of the redwood forest, my former wife, Norma, and I built a small second home, exquisitely designed by Gerald McCue, then chairman of the architecture department at the University of California. The setting was so private that one could reach it only by descending, through a screen of bay laurel, live oak, and fir, some 75 irregular steps made from railroad ties. (A dozen years later, in my post-divorce bachelor days, when I ended up in a full leg cast and crutches for over eight months as a consequence of a hiking accident, the 75 steps created an insurmountable barrier that drove me out of the house. I never resumed full-time residency in it thereafter.)

Around the same time, my son, not yet 20, was the beneficiary of a trust fund based on an early gift in Syntex shares, which had multiplied many times in value. He requested that the bulk of the trust be used to construct his own home—built in the shape of a hawk—near a small lake on the western portion of the ranch, an hour's hike from my ranch abode. In 1974, my daughter Pamela—then living in La Jolla while her husband was finishing medical school—followed suit. Her home and studio went up on the west side of Bear Gulch Road, half an hour's walk from her brother's home. Sometimes, when she sat on one of her hills overlooking the Pacific, and the ocean wind blew the right way, she could faintly hear Neil Young rehearse.

Pami and her husband had moved into their ranch home when the latter started his radiology residency at Stanford, where they had met as undergraduates. In July 1978, I'd been divorced for nearly two years and was also living at SMIP in the redwood house. For well over a decade, we had hiked weekends, and frequently on weekday evenings, all over the property. Still, there were many areas we'd never explored; some sections were simply too rugged or otherwise inaccessible. As I already recounted in an earlier chapter, in 1983, on one such hike with a group of my students from Stanford, I fell while climbing over some fallen logs at the bottom of a steep canyon. By the time I hit the ground my stiff leg was badly broken. If my students had not been with me, who knows how long I would have lain there before my body had been found? But I survived, whereas my daughter was already dead by then.

Many a death—especially when caused by accident or sudden disease—is met by the bitter question 'Why?' Invariably it is addressed to God or against God, meaning 'Why did You let that happen?' But there is another 'Why?' after a suicide—the cause of my daughter's death—which must be addressed to the person who is now dead. I was too depressed to ask at the time, nor was I ready to ask myself whether I could have done something to prevent it. On 4 July 1978, the day before she killed herself, Pami had hiked over to my home to spend a few hours with me in the sun, talking about her future. Nothing in her tone or conversation had given me any inkling that she was teetering on the edge of the precipice.

My immediate response to Pami's death was typical of how I coped at that period of my life with personal disaster: I drowned myself in work. Seventeen-hour workdays ensured that when I finally dropped into bed I fell immediately asleep. In addition, there was the legal and accounting work involved in being the executor of Pami's estate, which, like Dale's, had multiplied manifold with the rise of Syntex stock from the trust I had established in her infancy. But after 11 weeks of such work-induced anesthesia, I suddenly decided to travel, and invited my then-companion and now wife, Diane Middlebrook, to join me on a trip to Venice and Florence. I wanted to focus on the art and do it with the right partner.

Three evenings in a row, Diane and I sat in an outdoor café on the Piazza della Signoria facing the Palazzo Vecchio, to relive the day's impressions—and to talk about Pami's decision. Was it an inevitable consequence of a person suffering from depression who had been unwilling to consider therapy? Was it the chronic physical back pain that prevented her during the last two years of her life from doing any of the garden and animal work she loved? She could barely feed the horses that had meant so much to her, let alone ride them. Was it her disillusionment with the commercial art scene, with the humiliating compromises a young artist is called upon to make? Or was it the lack of professional peers resulting from her self-imposed isolation in the majestic but also overpowering setting of SMIP? Her husband, surrounded every day by a multitude of people in the hospital, had hardly a minute free for contemplation. Like me, he felt calmed upon returning in the evening to the solitude of the coastal mountains, often shrouded in the veil of the ocean fog pouring in through the canyons. The extraordinary silence was a soothing contrast to the cacophony of the workplace. But what about the person who remained behind all day? Does beauty in nature, when experienced in solitude, always calm and please, or does it also terrify? Some of Pami's bitterest poetry (published posthumously in chapbook form) was written during that period of her life.

Pami always read, and during those last few years she read a great deal. Much of it was women's writings and feminist literature, which provided fertile background to her ideas about the place of women in art. All her artistic role models, all her former teachers at the San Francisco Art Institute and at Stanford University, had been men. Indeed, until the time of her death, Stanford's art department had not a single woman faculty member among its studio art faculty. Diane, then head of Stanford's Center for Research on Women (CROW) in addition to her professorship in English, taught me a great deal about the daily affronts women sense in a male-oriented culture. I am sure that my own quasi-patriarchal style contributed to Pami's sensitivity. These, and many other issues, were the subjects of our evening talks in the Piazza della Signoria, practically at the feet of the prancing bronze horse on which Cosimo de' Medici surveyed the splendor his family had sponsored. 'It's hard to think what Florence would've been like without the Medicis,' I mused, pointing to the Giambologna bronze. 'But imagine what it would be today if their patronage had extended to women,' said Diane.

In that hour, my own response to Pami's suicide finally took shape. Suicide is a message to the survivors, but the text must be read by each individual—parent, sibling, spouse, friend—in the light of his or her past relationship with the deceased. An answer to the question 'why?' can be provided only by the survivor; on that September evening in Florence, I decided that my answer, or at least my response, would be patronage of the kind that would have benefited Pami. Diane became my partner—intellectual and operational—in this endeavor. In the 1960s, when we acquired the land that became SMIP, my children agreed with me that it should be kept in an unspoiled state for the ultimate benefit of the public. That decision contributed to the legal establishment of the 'Djerassi Foundation,' a nonprofit entity that we envisaged as the eventual beneficiary of our respective testaments. Whatever philanthropic donations I made during the late 1960s and 1970s were funneled through that foundation, but any substantial activities were meant to await my death. In 1978, Pamela and Dale were the two trustees empowered to decide how my own estate—land, art, and remaining assets—all derived indirectly from the Pill, would be distributed by the Foundation; but none of us was then thinking of death, and the full operation of the Foundation still seemed many years away.

All this changed on 5 July 1978, when a bottle of pills transformed the concept of the Djerassi Foundation into a real entity with substantial financial resources and title to a significant portion of SMIP ranch land as well as to Pamela's house and studio. Over the course of the next few years, I augmented the Foundation's endowment by donating most of my SMIP property with the exception of my ranch home, together with additional funds arising from the sale of my substantial collection of world-famous dead artists. My daughter's suicide and the financial resources derived from the Pill via my Syntex stock options had thus led me, a modest patron of the Arts, to the much more significant patronage of artists in the broadest sense of the word. Even the name of the property had changed.

Visitors had often asked what SMIP meant. Instead of giving the simple, if corny, answer, 'Syntex made it possible', I usually hedged with the reply, 'guess.' Depending on the respondent as well as the occasion, the guesses were extraordinarily wide ranging, convincing me that S M I P is clearly the most promiscuous array of acronymic letters. 'See me in private' or 'sexy man invents pill' were one extreme of a continuum ending with 'surely mankind is precious' on the elegant side, with 'spend my investment portfolio' or 'some men ignore promises' in between. At one stage, I had counted over 40 variants. But the most appropriate—and in the end the one that stuck and now greets visitors on a moss-covered sign by the entrance—is *Sic manebimus in pace* (thus we shall remain in peace). Its originator was the physicist Felix Bloch, Stanford's first Nobel laureate, who produced it within 30 seconds after my challenge. This was during the height of the Vietnam War, when open opposition to that war landed me on Richard Nixon's White House 'Enemy list,' and the question of peace was on our minds.

V

That autumn of 1978, when Diane challenged me in Florence about patronage of women artists, I decided to offer to Stanford University's art department the use of my daughter's ranch home and studio as well as an annual stipend to underwrite one-year residencies at SMIP for women artists of accomplishment. The incumbent would have no formal teaching duties, but should be prepared for some open-studio events and for interaction with art students and faculty. Toward the end of her stay, the artist would present an exhibition of the past year's work at the Art Museum of Stanford University. I felt that this would accomplish at least two objectives dear to Pamela's heart: patronage through a public exhibition in a museum, free from commercial considerations; and exposure of mature women artists to the Stanford art community, which had plenty of women at the student level but none among the studio faculty. But while the museum director agreed to the stipulation for an exhibition, and some art history professors, notably the late Albert Elsen, were supportive, the studio art faculty was not prepared for any obligations that would make such a visiting artist feel welcome. In the end, CROW, the Center for Research on Women then headed by Diane, became the Stanford sponsor, and members of CROW and some outside art professionals and museum directors formed a selection committee that began to canvass nominators all over the world.

After operating in this mode for four years, Diane and I finally concluded that, although the driving distance from Pami's studio to Stanford University was only a few miles, the attitudinal gulf was simply too wide. The women artists got a lot done under close to ideal physical conditions, but their presence had only a minimal effect on the Stanford art scene. We decided that the buildings owned by me on my side of SMIP ranch—the ranch manager's residence and the 12-sided barn—would lend themselves to conversion into a small artist's colony, which would overcome the sense of isolation and lack of peer interaction (also Pami's problem) encountered by the artists who had come alone for an entire year. I began to realize that free creative time without conditions may be the ultimate gift for an artist, but it is an incomplete gift if it denies the crucial human need: companionship.

Diane, in her capacity as a Djerassi Foundation trustee, visited the two oldest artist's colonies in the East, Yaddo and McDowell, as well as two smaller ones, the Edna St Vincent Millay Colony and Hand Hollow—the latter founded by a friend of mine, the kinetic sculptor George Rickey, on his farm in East Chatham, New York. Diane's report and Rickey's counsel persuaded me to convert the former manager's four-bedroom residence into an eight-bedroom, five-bath house and to create two studios in the barn. By late 1982, the Djerassi Foundation artist's colony was born. The gender requirement was dropped as the community expanded; and to make our program available to a larger number of artists, residencies were limited to periods of one to three months. Since all the bedrooms would be large and, with one exception, have balconies or direct access to the surrounding garden, they could easily double as workrooms for the writers. Visual artists and composers would use the two studios in the barn. Rickey considered good food to be indispensable to the success of an artist's colony—advice we followed by hiring his chef, who had cooked at Hand Hollow with *cordon bleu* elegance.

By the end of that first year, we had housed and fed 52 artists, 28 of them women. Over the years, the friendships and collaborations formed at SMIP extended beyond the professional to at least three marriages (the sculptors Patricia Leighton from Scotland and Del Geist from New York; the Houston painter Josefa Vaughan and the San Francisco composer Charles Boone; the literary critic, NPR broadcaster and author, Allan Cheuse and the dancer Kriss O'Shee) and any number of short- or long-term liaisons. Our first composer, John Adams, who later gained international recognition for his opera *Nixon in China*, spent three months at SMIP, where he composed the music to *Available Light*, a work commissioned from him, the New York choreographer Lucinda Childs, and the Los Angeles architect Frank O. Gehry for the opening of the Museum of Contemporary Art in Los Angeles. On several occasions, Adams remarked to me how difficult it was to collaborate with his choreographer, Lucinda Childs, three thousand miles away. His comments convinced me that we should encourage interdisciplinary collaboration by creating additional studio spaces within the twelve-sided barn, dedicated to choreography and performing arts, to music, to photography, and to ceramics. This expansion we eventually realized through financial assistance from the James Irvine Foundation, the William and Flora Hewlett Foundation, and somewhat later the John D. and Catherine T. MacArthur Foundation.

The choreography space—by far the largest studio—has also served as the site for performances suitable for nearly one hundred spectators. Everything from premier dance, theater, or music pieces has been presented over the years to supporters and visitors of the Program. For instance, John Adams, who subsequently became a trustee, presented his opera, *Nixon in China*, in video and lecture form in that space.

At the end of our twentieth year, the program has grown to such an extent that, with over 1000 artists from 38 states and 29 foreign countries, it has become one of the two largest resident artist's colonies west of the Mississippi. The artists have included a Nobel Prize laureate, some MacArthur Fellows, numerous Guggenheim fellowship holders, and winners of literary and visual art awards, as well as many artists who had not yet received wide public recognition but were considered worthy of support by our selection panels. It has long passed the stage of a private family foundation to become a totally independent non-profit entity, renamed the Djerassi Resident Artists Program, run by its own board of trustees, and supported by many government agencies, private foundations, and individual donors.

VI

With one exception, there is no requirement for any artist to leave any portion of completed work to the Program at the end of the residency. The exception was prompted by Paul Klee's guest book, which his son Felix had shown to me in Bern. Every artist, irrespective of discipline, is presented with a single, large sheet of drawing paper and requested to leave behind some artistic or intellectual statement. The collection of hundreds of such sheets constitutes the most personal of the many other forms of archival documentation (photographs, videos, tapes) that the Program has assembled. Two public exhibitions have already resulted from this ever-expanding collection. One was a selection of artist's pages by composers, selected by Charles Amirkhanian (himself a composer and a former director of the Program) and exhibited in the foyer of the San Francisco Symphony Hall for many weeks during one of its concert seasons. The other was an ambitious selection of drawings curated by the present director, the sculptor and former museum director Dennis O'Leary, which became the 1999 summer show at the Palo Alto Art Center.

Around the twentieth anniversary of the Program, the Peninsula Open Space District acquired in perpetuity all development rights to the SMIP ranch property—an idea spawned by my son, Dale, the only family member now on the Program's board of trustees. Amounting to nearly 2 million dollars, this largest outside financial contribution to the Program now guarantees that the spectacular environs of SMIP can never be altered, while at the same time substantially increasing the Program's endowment, thus assuring its continuing operation. A line from the artist's page, *Landscape near Bear Gulch Road*, by Janet Lewis, who at 90 was the oldest artist-in-residence, expresses with poetic sparsity the grandeur of that natural setting:

> One tone, one visible substance, a splendor
> Of multiplicity and self-abandon,
> Beneath that band of intense blue
> Lacking which, the hill is incomplete.

And on a more intimate level, so do these lines from Amy Clampitt's page, entitled *Coming upon an unidentified work of outdoor art near Woodside, California.*

> The sudden, gladed glimpse of something made.
> Whose doings are these stationings
> of jute and brushwood, these totem—
> tripod harp shapes, that blood-red lyre whose
> silence stuns, among the redwoods—a holy
> place, an offering, a condition harking
> back to states of soul (an Eden's trespass
> slyly trailing withes of poison oak)
> one gropes to ascertain the name for?

Clampitt was referring to a kind of art that has transformed SMIP into an outdoor sculpture park, albeit for hardy devotees, since it takes easily four hours of strenuous hiking to view it all. Numerous environmental sculptors have used the artistic output of their residency to donate to the Djerassi Resident Artists Program site-specific work of great esthetic and even financial value. One of them is David Nash, one of the most distinguished past artists-in-residence. A British sculptor now working in Wales, he first came to the Program in 1987 at the time of his retrospective exhibition at the San Francisco Museum of Modern Art. While wood is his sole medium, and chain saw or ax his principal tools, he had never before handled redwood or madrone, the two dominant species in our forest. During his first stay, he had created a group of madrone sculptures for a highly successful show in Los Angeles; in addition, out of a huge redwood trunk that had lain for decades in Harrington Creek, Nash had fashioned *Sylvan Steps*—a Jacob's ladder rising at a steep angle out of the water into the sky.

When he first selected that site—accessible only along the creek bed by clambering over rocks and fallen timber—he was unaware that in a small waterfall a few hundred feet upstream we had, in 1978, scattered my daughter's ashes. Within a minute, some flecks of them must have floated past that sculpture site.

Two years later, during his second residency, on Thanksgiving Day, Nash and I had already been hiking for several hours in search for some felled redwood trunks at least five feet in diameter. It was the minimum size he required for the three-part sculpture he intended to site around some of the burned-out giant redwood stumps that can still be found, here or there, on our property from nineteenth-century logging. We had located four sites in the forest where blackened trunks rise out of the bracken—just the right backdrop for the scorched pyramid, cube, and ball Nash planned to shape, but still missing was the right arboreal progenitor for these forms. Of course, we crossed the shadow of many a living redwood giant, but cutting one of these was out of the question. That's when I recalled that some selective logging had just been completed on our neighbor's land across Bear Gulch Road; only a few days ago, I had followed impatiently a slow-moving truck stacked high with redwood logs. Perhaps 'our' piece had not yet been removed.

We did not expect anyone to be working there on Thanksgiving Day. But, after climbing over the locked gate and walking down the forest road, inches deep in dust (it had not rained for weeks), we heard in the distance the grinding of gears. Soon we came upon a mammoth tractor setting up erosion breaks to preserve the roadbed during the winter rainy season. 'Have you moved out all the logs?' I shouted up to the bearded driver after he had shut off the thundering engine. 'We need . . .,' I said, and then explained who David Nash is, and why we were searching for a special fallen redwood rather than a turkey for Thanksgiving. 'All gone,' he said, but then remembered. 'A big one fell across the fence near the property line. Probably years ago . . . in some storm.' According to him, it was partly rotten—sufficiently so that it had not been worthwhile to haul it to the mill. Nash was dubious that it would do, but I said, 'Let's look anyway.'

We followed the man's directions to the fence a half-mile down the logging road. When we finally came upon it, I was dumbfounded. Eleven years ago, I had hobbled here as fast as my stiff leg would carry me—but from the opposite direction, down the meadow from our side of the property, toward this fence across which the massive trunk was now lying, broken into three enormous pieces. It was the spot where my daughter had killed herself, where I had never dared to return. We found the rot to be only superficial; the wood was precisely what David Nash had been seeking all Thanksgiving long.

VII

At one time or another, I have met many of the artists for evening meals around the large dinner table—occasions that are usually followed by readings, slide presentations, music or dance performances, or just good conversation. As I listen to music just composed, or face a canvas, still wet on a studio wall; as I hear a poet read lines that could only have been born here; as I look with a sculptor for a site, in the forest where once I searched for Pami's body; I catch myself wondering what she would have thought of all this.

Five years after Pami's death, I received a panicky phone call from an artist then occupying my daughter's former home and studio, saying she had found a note, deep in back of a drawer. She was shaking when she handed me the piece of paper with Pami's handwriting on it, a ghost's message. That night I recorded this in my journal:

> My only daughter,
> I find this note:
>
> 'I have nothing left to say,
> So l don't talk.
> I have nothing left to do,
> So I close up shop.'
>
> No date
> No address
> No signature
> Your handwriting.
>
> Written for whom?
> Yourself?
> To whom it may concern?
>
> Written when?
> Days,
> Weeks,
> Perhaps months
> Before you walked into the woods?
>
> If only you'd said these words to me.

If only, I add in retrospect, your death had not been necessary before I took seriously the patronage of the living.

Chapter 11

Science on stage

By now, readers may well wonder what other personal transformations of mine I credit to or blame on the Pill. There is one more: playwriting. It is my most recent foray into another creative field and, given my age, may prove to be the last. I have been a theater aficionado all my adult life, but for the past two decades, the theater has been my most captivating avocation and also my most frequent mode of relaxation. I attend 30–40 plays a year, many of them during my summers in London, where, since the 1980s, I have gone to write fiction. I took up writing with a very specific aim in mind: to use the novel to transmit specific information. This is, after all, the reason scientists publish their scientific work. But can this be sufficient justification for a scientist to publish fiction and now even plays?

In 1990, one of my favorite novelists, David Lodge, published an insightful essay, *The Novel as Communication*, in which he ruminates on objections raised by distinguished critics such as Roland Barthes and Walter Benjamin to fiction written from such motives. Yet Lodge sensibly concluded that 'most writers and readers of fiction outside the academy, it must be said, still subscribe to the communication model of the literary text.' Some of his novels, such as *Small World* or *Nice Work*, are prime examples how information can, in fact, be presented in the guise of fiction. And why is Lodge so successful? Because in my judgment, he clearly follows Horace's famous prescription from *Ars Poetica*: '*Lectorem delectando pariterque monendo*' [delighting the reader at the same time as instructing him]. I figured that what was good enough for Quintus Horatius Flaccus, and 2000 years later for David Lodge, ought to be kosher for Carl Djerassi. But after having launched five novels with that stated purpose openly announced in forewords, why have I now switched—at least temporarily—to play writing with a similar objective?

I

Science is inherently dramatic—at least in the opinion of scientists—because it deals with the new and unexpected. But does it follow that scientists are dramatic personae, or that science can become the stuff of drama? Or do our idiosyncrasies appear so queer or dull to the rest of the world that we will never appear in successful plays unless represented in the extreme? Until now, 'science-in-theater' has been a rare genre, although playwrights of the caliber of Brecht, Dürrenmatt, Whitemore, and Stoppard have on occasion chosen scientists or scientific themes as components for the plots of major plays.

A more recent phenomenon on the London theater scene is the appearance of 'pure' science-in-theater plays by prominent playwrights who are not scientists. Steven Poliakoff's *Blinded by the Sun* attempts to illuminate through a theatrical version of the chemical 'cold fusion' debacle of the early 1990s some of the idiosyncratic aspects of a scientist's drive for name recognition, as well as the competitive aspects of a collegial enterprise. Michael Frayn's *Copenhagen* calls upon quantum mechanics and the uncertainty principle for much of the scintillating interplay during a wartime encounter in Copenhagen between two physicists, Werner Heisenberg and Niels Bohr, under the skeptical eye of Bohr's wife. Although Frayn made no concession to scientific illiteracy, the play still became a major West End as well as Broadway theatrical success. In France, Jean-Noël Fenwick's play, *Les Palmes de M. Schutz*, dealing with Marie and Pierre Curie's discovery of radium in a realistic stage replica of a laboratory, went way beyond a *succes d'estime* to be turned into a film. The Italian title of the play, *Amore e chimica*, though kitchy, even dared to bring the word 'chemistry' into the title.

But the success of a few isolated incidents still leaves open the question: can 'science-in-theater' also fulfill an effective pedagogic function on the stage, or are pedagogy and drama antithetical? Must the urge to education be an automatic kiss of death when writing for the commercial theater? 'Didactic,' which is code for 'boring,' is usually the most damning term a reviewer can use to drive a prospective audience from a play. Given such a prejudice, especially in the make-or-break context of theater reviewing, any playwright openly admitting to such a didactic impulse risks accusations not only of didacticism, but also of a masochistic desire for instant infanticide of a play barely written.

Yet I still remember the evening in September 1996 when I turned to my wife, as we walked out of the Royal National's Cottesloe Theater, to announce that I was going to write a play myself. It was a decision prompted by the mixed feelings the play we had just seen—*Blinded by the Sun*—had engendered in me as a scientist. As the first act had progressed, I was impressed by how accurately a non-scientist had caught some of the behavioral characteristics of academic scientists. There was also the precise evocation of the atmosphere of a 'red-brick' university in the UK—so different from the California academic hothouse I inhabit. (At that point, I did not know that Poliakoff's brother was a chemistry professor.) He even negotiated successfully that most perilously didactic shoal, where so many scientific narratives come to grief, the patient explication to a conveniently uninformed character of the scientific problem under investigation:

> Water contains hydrogen. But how to get it out? Some chemical reactions are caused by shining a light. Find the right chemical to act as a catalyst—shine a light, a beam, above all the sun—and you can create hydrogen out of sunlight and water. Hydrogen, which will run planes, cars, anything you want. And when you burn it, it will turn back into water. Polluting nothing.

So far, so good. After all, that's exactly why the initial claims of cold fusion the electrochemists Pons and Fleischmann made in 1989 had caused such a sensation among chemists and physicists and even the general press. But a few minutes later, scientific didacticism raises its ugly head.

Elinor

Is it anatase or rutile? You haven't used an adsorbed dye to shift the Lambda-max, clearly—

Christopher

The particles have an electrodeposited coating. It's only a few nanometers thick so refractive index matching makes it—

Elinor

Yes, it certainly seems to have a high quantum yield. Maybe there's an added sulfonated surfactant to enhance mass transport at the surface?

Christopher

No. Think more of a catalytic system—

To an audience equipped with more curiosity than knowledge, such lines can be only meaningless gobbledygook. Did Poliakoff have his scientists speak them because he felt that a theater audience would not have understood the real chemistry? Or is the playwright, dealing with serious science, faced by an intrinsic barrier no amount of dramaturgy can conjure away?

Not too long ago, on 30 June 2000, I explored such questions of science-in-theater in the inaugural Dennis Rosen Memorial Lecture at the Royal Institution in London. For the past 200 years, since the days of Humphrey Davy and Michael Faraday, its acoustically perfect amphitheater has been the science theater *par excellence* of the London scene, but a very special theater with only a single actor—the scientific lecturer—and only a single subject, science. Monodrama is not unheard of in theater, but the bulk of Western drama since its origins in Greece has emphasized dialogue and human interaction among actors. Even the audience of this special science theater, the Royal

Institution, is unusual. People come here for a specific purpose: to be instructed about matters scientific, and preferably to be entertained as well. For example, the poet Samuel Taylor Coleridge came to Davy's scientific lectures at the Royal Institution to 'increase his stock of metaphors.' But I came to the Royal Institution not to talk about bringing theater to science, but rather science to the theater; to answer the question, is 'science-in-theater' a viable genre, or is the phrase itself a contradiction in terms?

As I have said on more than one occasion, it is the conviction of many scientifically untrained persons that they are unable to comprehend scientific concepts that stops them from even trying. For such an audience, rather than an unadorned lecture, 'case histories' can be a more alluring as well as more persuasive way of overcoming such hurdles. If such a story-telling 'case history' approach is employed on the stage rather than on the lectern or the printed page, then we are dealing with 'science-in-theater.'

But not all audiences (as the *New York Times* critic Bruce Weber once wrote) are so conditioned by low-brow entertainment that they are prepared only to have their senses tickled, not their brains massaged. Of course, most non-scientist playwrights use science for metaphorical purposes. I rather doubt that Tom Stoppard's motivation in writing *Hapgood* was to illustrate Einstein's photoelectric effect or Heisenberg's Uncertainty Principle, both of which are described at length by a physicist-turned spy, named Kerner. Here, for instance, is a brief excerpt from one of Kerner's speeches:

> The particle world is the dream world of the intelligence officer. An electron can be here or there at the same moment. You can choose; it can go from here to there without going in between; it can pass through two doors at the same time, or from one door to another by a path which is there for all to see until someone looks, and then the act of looking has made it take a different path. Its movements cannot be anticipated because it has no reasons. It defeats surveillance because when you know what it's doing you can't be certain where it is, and when you know where it is you can't be certain what it's doing: Heisenberg's uncertainty principle; and this is not because you're not looking carefully enough, it is because there is no such thing as an electron with a definite position and a definite momentum . . .

Stoppard was writing a fiendishly clever whodunit—not an explication of twentieth century physics—and he was neither the first nor the last to use Heisenberg's physics as metaphor. Whatever else his Kerner demonstrates here, his speech suggests that audiences are not so allergic to science *per se*.

But what of theater in which the science is not exploited for its metaphoric possibilities, but becomes its central subject: is such a theater possible? In the days before Michael Frayn's *Copenhagen* became a hit, whenever I decried the dearth of 'science-in-theater,' three plays were invariably brought up as examples of that genre: Brecht's *Life of Galileo*, Dürrenmatt's *The Physicists*, and Tom Stoppard's *Arcadia*. But Brecht's and Dürrenmatt's motivation was primarily to express their skepticism about science; the actual science played a minimal role. Brecht's politics made him question any science that was not devoted to the service of the people, while Dürrenmatt, in expressing his fear of nuclear annihilation at the height of the

Cold War, put his Newton, Einstein and Möbius characters into an insane asylum as his metaphor for the physicist's world. Brecht's *Galileo*, of course, illuminates the conflict between religion and science and the ultimately flawed natures of scientists and of men of the cloth—topics that for me make that play much more timely than *The Physicists*. A 1999 production of Dürrenmatt's play that I saw in Vienna's Volkstheater seemed to me totally outdated in its message.

That leaves *Arcadia*. I am an enormous fan of Stoppard's plays, so much so that my admiration borders on uncritical idealization. *Arcadia* is a close second to my personal favorite, *Travesties*. But is *Arcadia* 'science-in-theater'? Of course it has didactic sequences—indeed rather long ones. A description of Fermat's last theorem appears within the first few minutes of the play, right after a long and hilarious definition of 'carnal embrace.' Septimus Hodge, the tutor, has this to say:

> Carnal embrace is sexual congress, which is the insertion of the male genital organ into the female genital organ for the purposes of procreation and pleasure. Fermat's last theorem, by contrast, asserts that when x, y and z are whole numbers each raised to power of n, the sum of the first two can never equal the third when n is greater than 2.

But while Fermat reappears here and there for brief moments, it is iterated algorithms and chaos theory that occupy the better part of a scene in *Arcadia*, with one monologue covering almost an entire page of text. Some of it is quite straightforward:

> You have some x- and y-equation. Any value for x gives you a value for y. So you put a dot where it's right for both x and y. Then you take the next value for x which gives you another

> value for y, and when you've done that a few times you join up the dots and that's your graph of whatever the equation is . . . [But] what she's doing is, every time she works out a value for y, she's using that as her next value for x. And so on. Like a feedback. She's feeding the solution back into the equation, and then solving it again. Iteration, you see.

Later sequences become more complicated, as they explain the use of such approaches to contemporary population biology. But I suspect that Stoppard's real motivation for many of these factual forays was not to teach his audience about iterated algorithms. Rather, *Arcadia* has to do with nature and how we humans handle and mishandle, or understand and misunderstand it. Take this passage from the same scene:

> People were talking about the end of physics. Relativity and quantum looked as if they were going to clean out the whole problem between them. A theory of everything. But they only explained the very big and the very small. The universe, the elementary particles. The ordinary-sized stuff which is our lives, the things people write poetry about—clouds—daffodils—waterfalls—and what happens in a cup of coffee when the cream goes in—these things are full of mystery, as mysterious to us as the heavens were for the Greeks. We're better at predicting events at the edge of the galaxy or inside the nucleus of an atom than whether it'll rain on auntie's garden party three Sundays from now.

A few lines later comes the punch line, 'It's the best possible time to be alive, when almost everything you thought you knew is wrong.' A statement which, with its ironic edge, brings Stoppard closer to a Brechtian skepticism than it brings its audience to an appreciation of science—a word that means, after all, 'knowledge.' For all its scientific literacy, the science in Stoppard's plays—even the most didactic sequences—is there because Stoppard decided to write a play for which scientific concepts are useful and intellectually attractive metaphors. But they are not intrinsic to the story. *Arcadia* could have been written without Fermat's last theorem, or even without chaos theory, simply as a literary lark around the historical figure of Lord Byron. It would not have been the same play—and to me a less interesting one—but it still would have been performed and most likely also acclaimed as a successful play.

This is less true in Hugh Whitemore's *Breaking the Code.* Fairly early in the first act of that marvelous play, in Scene 5, the mathematician Alan Turing delivers a monologue three pages long, which any instructor in Playwriting 101 would consider grounds for dropping the student from the class. Yet Derek Jacobi, in the role of Alan Turing, did manage not only to describe David Hilbert's insistence on consistency, completeness, and decidability as basic requirements for mathematics in an elegantly accessible manner, he goes on just as engagingly to summarize Kurt Gödel's subsequent demonstration that no mathematical system could be both consistent and complete. You will have to take my word for it—without my copying three pages of text—that this extremely didactic prose works even for audiences that had never heard of Hilbert or Gödel—two of the brightest stars in the mathematical firmament. Why did Whitemore put them there? Because the theme of his play was the 'Turing machine' and its use in the breaking of the German Enigma code; and to understand that, some of the underlying math was indispensable. But Whitemore goes farther than the brute narrative requirements of his situation: he uses the same mathematical didacticism for a touching interlude between Turing and his female friend Pat Green where Turing produces a pine cone and tells her about Fibonacci sequences:

> A Fibonacci sequence is a sequence of numbers where each is the sum of the previous two; you start with one and one—then one plus one equals two—one and two, three—two and three, five—three and five, eight—five and eight, thirteen . . . Now look at that fir cone. Look at the pattern of the bracts—the leaves. Follow them spiraling round the cone: eight lines twisting round to the left, thirteen twisting to the right. The numbers always come from the Fibonacci sequence . . . And it's not just fir cones—the petals of most flowers grow in the same way . . . And it prompts the age-old question: is God a mathematician?

Of course, it is the very next line that demonstrates that we are dealing with human interactions in a drama and not just a lecture, as Pat replies: 'I love you, Prof. I love you. You know that.'

But what about the rare *pure* 'science-in-theater' play like Michael Frayn's *Copenhagen*? Though pleased by what the success of this play has done for 'science-in-theater' and the sudden attention that genre (rather than just this specific play) is now starting to receive (just consider a series of major articles on that topic by no less than three different critics in the 9, 12, and 14 April 2000 issues of the *New York Times* alone), I am still surprised at *Copenhagen*'s meteoric rise. That Frayn was an established and highly skilled playwright known for his humor undoubtedly helped. Yet I am convinced that if the identical script were thrown over the transom by an unknown playwright it would not even have been read, let alone produced by the same London or Manhattan theaters that actually staged *Copenhagen.* Pages of didactic exposition by two characters, where uncertainty in perception and memory rather than dramatic excitement reigns, is not the stuff of which hits are usually made. I loved the play—and have seen it twice—at the Cottesloe and in the West End. But what about the bulk of the audiences—the non-scientists? Do they perhaps accept such material more in a spirit of awe, similar to the one that made several million people buy copies of Stephen Hawking's *Brief History of Time* for display on coffee tables rather than because they had suddenly become fascinated with the intricacies of cosmology? But how would the average literary manager of a theater respond if he had happened on the following page of a script by an unknown playwright?

Heisenberg

Max Born and Pascual Jordan in Göttingen.

Bohr

Yes, but Schrödinger in Zürich, Fermi in Rome.

Heisenberg

Chadwick and Dirac in England.

Bohr

Joliot and de Broglie in Paris.

Heisenberg

Gamow and Landau in Russia . . .

Bohr

You remember when Goudsmit and Uhlenbeck did spin?

Heisenberg

There's this one last variable in the quantum state of the atom that no one can make sense of. The last hurdle . . .

Bohr

Pauli and Stern are waiting on the platform to ask me what I think about spin . . .

Heisenberg

Then the train pulls into Leiden.

Bohr

And I'm met at the barrier by Einstein and Ehrenfest.

Of course it is totally unfair to quote excerpted passages from a play out of context, but the above may well represent a world record for the number of different surnames—sixteen in all—appearing on a single page of a script. What seems most significant to me about such passages is that, in them, we find not science as a metaphor for human affairs, but human affairs functioning as metaphors for scientific phenomena. Only in a play functioning successfully both as theater *and* as science education could such a trope engage an audience's attention. It certainly could not have been for the euphony created by this potpourri of European surnames, many of them familiar only to physicists.

To my knowledge, *Copenhagen* received uniformly complimentary reviews and commentary with one exception: a serious critique by an American historian, Paul L. Rose, the author of a scholarly book entitled *Heisenberg and the Nazi Atomic Bomb Project.* In his lengthy piece in the *Chronicle of Higher Education*, Rose compliments Frayn on the theatrical aspects of *Copenhagen*, but chastises him severely on a major didactic point. Not on quantum physics and complimentarity, but on the revisionist nature of Frayn's interpretation of Heisenberg's role in the putative German atomic bomb project. Frayn anticipated this argument, which is why his published text of the play contains a densely written postscript defending his interpretation of the historical record. Nevertheless, this controversy illustrates one hazard for the playwright brave enough to venture on factual terrain. Not only does one run the usual risk of the theater critic's sharp stiletto, but also the danger of raising the expert's hackles regarding a perceived error.

II

To write science-in-theater does not require that the author be a scientist. All the plays I have mentioned so far were written by internationally recognized playwrights, who had gained their scientific knowledge second-hand. Still, why is it that so few 'hard' scientists—and no chemists at all, as far as I am aware—have become recognized playwrights, whereas some physicians have made major contributions? Consider Anton Chekov or Arthur Schnitzler, for instance. Is it because chemists find it difficult to communicate without recourse to blackboard or slides or some other kind of pictogram? Or is it because chemists deal primarily with abstractions at the molecular level, whereas physicians spend their days listening to the stories of other human beings? Even the most scientifically-invested plays succeed, if they do, because they work at the human level. Or is it that all formal written discourse of scientists is always monologist, whereas the theater is the realm of dialogue?

Perhaps none of those generalizations is the reason, yet it is that last one that tempts me the most, especially when I consider my own forays on the stage. I find dialogue so stimulating, both as audience and speaker, that an exceptionally high proportion of my fiction takes place in dialogue rather than narrative. I first became intrigued by the idea of 'science-in-theater' when I was working on the third installment of my science-in-fiction tetralogy, *Menachem's Seed.* The science of that novel involved reproductive biology as seen by a childless woman scientist preoccupied with childbearing. Much of the factual information of that novel concerns the current high-priority area of assisted reproduction and the reasons why research on contraception is moving off the main stage.

While I recognized at the outset the inherent problems of converting any novel into a play, it seemed to me that building a science drama based on my novel's scientific themes and human conflicts was not unreasonable. In addition, reproduction is inherently more dramatic than any abstract chemical concept. At least, those were my thoughts after I left the Cottesloe Theater at the conclusion of Poliakoff's *Blinded by the Sun*; as we headed home, I found myself starting to think of the title of my first play. When I write, I start toying with the title the moment I have settled on the theme—before I have even worked out the plot. The title *Menachem's Seed* was too suggestive of a dramatic adaptation of my novel. 'ICSI,' though describing factually the reproductive technology I wanted to feature in my play, offered no subliminal association for a prospective theater audience. 'Condom Capers' implied a comedy and 'The Purloined Sperm' a mystery. *An Immaculate Misconception* finally won out, even though the double meaning of 'misconception' is lost in languages (such as French) into which the play's title has now been translated verbatim. In two other productions, the title was altered: the German *Unbefleckt* or the Swedish *Obefläckad*, both standing simply for 'Immaculate,' at least raises an attractive element of ambiguity.

Unlike the playwrights I have mentioned so far, who mostly used science for theatrical aims, I started from the opposite side, using the stage for a scientific (or at least pedagogic) purpose. But even though we start at opposite ends, for any play to work, we must meet in the middle: on the stage facing a live audience that did not come to be educated, but rather entertained. The sub-title of my play, *Sex in an Age of Mechanical Reproduction*, is an allusion to Walter Benjamin's famous essay of 1936 on *Art in an Age of Mechanical Reproduction*. I chose it because I consider the impending separation of sex (in bed) and fertilization (under the microscope) one of the fundamental issues facing humanity during the coming century. I picked Benjamin's phrase for a second reason as well: in our preoccupation to conceive, we often forget the product of all the technologies we utilize, namely the resulting child. Benjamin argues, '*The technique of reproduction detaches the reproduced object from the domain of tradition.*' All the reader has to do is to substitute 'child' for 'reproduced object' in order to land right in the middle of the ethical thicket that reproductive technologists invariably face: they support heroic efforts by many couples to overcome certain biological hurdles that may very well harm rather than benefit the 'reproduced object.'

For the didactic component of my play, I chose the most ethically charged reproductive technology of them all, ICSI. I suspect that few will argue with my assumption that everyone has opinions about reproduction and sex, and that most people of theater-going age are convinced that they know the facts of reproductive life. But do they really? I would offer odds that few in such an audience could answer correctly the following question: While it takes only a single sperm to fertilize an egg, how many sperm must a man ejaculate in order to be fertile? Answer: a fertile man ejaculates on the order of 100 million sperm during intercourse; a man ejaculating 1–3 million sperm—seemingly still a huge number—is functionally infertile. Ten years ago, there was no hope for such men. But now, many can become fathers because of ICSI. Yet how many members of the theater audience I wish to attract have heard of ICSI? How many of my audience here unless they had just read Chapter 5 of this book?

The 'scientific' way to find out would be to search the technical literature for the original publication by André van Steirteghem and collaborators in one of the 1992 issues of *Lancet* . Of my intended theater audience, few are likely to go to the medical literature, but this is what they would find:

> The ICSI procedure consists essentially of aspiration of a single spermatozoon from a droplet containing 10 per cent polyvinylpyrrolidone in buffered Earle's medium. The eggs are held by a holding pipette on the heated stage of a Nikon

> Diaphot inverted microscope at x400 magnification while the injection pipette is pushed through the zona pellucida into the cytoplasm . . . Even round-headed spermatozoa can be selected for ICSI, although normally they are unable to fertilize the egg owing to the lack of the acrosome.

Of course, many people will now go to the web rather than a science library. The patient web surfer might be led to Chapter 18 of my *Menachem's Seed.* What the reader of 'science-in-fiction' would find there would certainly be more comprehensible than the cold, impersonal prose of the typical scientific paper:

> I watched on a color video monitor connected to the microscope. I will never forget the moment when Van Steirteghem himself—in homage to me as director of the foundation supporting his work—immobilized an egg on the petri dish, with its polar body in the 12 o'clock position. I held my breath as he pushed the capillary needle, containing a single spermatozoon—aspirated tail-first into the injection pipette—through the zona pellucida from the 3 o'clock side. The sudden spasm I sensed in my lower abdomen as the tip of the needle stopped to eject the solitary sperm deep inside the egg gave way to tears as the needle was withdrawn with infinite gentleness and the injected egg released. Of course, I'm romanticizing the event, but it seemed I had witnessed a solemn event, primordial in nature.

The procedure, though conducted under the microscope, can be watched in real life on a video monitor connected to the microscope. In my *Immaculate Misconception*, the same scene is played out in front of the audience by projecting a *real* ICSI procedure on a large rear screen accompanied by the following dialog between the biologist, Dr Melanie Laidlaw (conducting the first ICSI procedure in history) and Dr Felix Frankenthaler, her clinical colleague, who will eventually transfer the fertilized egg back into the woman's uterus.

Melanie

[Puts on rubber gloves]

Would you start the VCR?

Felix

Sure.

[Pushes the button and turns toward the monitor so that only part of his face is seen. Both are completely silent as the screen lights up. MELANIE is hunched over the microscope, both hands manipulating the joysticks on each side of the microscope. She sits so as to be able to coordinate her words to action on the screen.]

Ah . . . here we are.

Melanie

[As single active sperm appears at bottom of image, excitedly interrupts]

Okay . . . let's see whether I can catch it with this capillary, one-tenth as thick as a human hair. But first I've got to crush its tail so the sperm can't get away . . .

[Quickly moves pipette toward sperm and sounds jubilant as the injection pipette crushes the sperm's tail]

Gotcha!

Felix

Not bad. Not bad at all.

Melanie

Now comes the tricky part. I've got to aspirate it tail first . . . As soon as I get close enough, just a little suction will do the trick . . . Hah! Gotcha!

[Screen image displays the sperm, tail first, being sucked into the pipette. Image now shows MELANIE 'playing' the sperm's head by moving it back and forward to demonstrate that she can manipulate it easily.]

Felix

Quit playing with him!

Melanie

I'm not playing with it. I just want to be sure that I can manipulate it at will. And why do you always call sperm 'him'? Is it because the sex of a baby is always determined by the sperm?

[Silence for a few seconds.]

Here we are.

Image of egg appears.]

Isn't she a beauty? Just look at her . . . here you are my precious baby . . . now stay still while I arrange you a bit . . . while I clasp you on my suction pipette . . . polar body on top . . .

[Frankenthaler points to polar body.]

Like a little head. I want it in the 12 o'clock position.

[Egg on screen is now immobilized in precisely the desired position for the penetration.]

Felix, now cross your fingers.

[He leans forward, clearly fascinated. Injection pipette containing sperm appears on image but pipette remains immobile.]

Felix

[Points to pipette on extreme right of image.]

What's the problem? It isn't usually that difficult.

Melanie

No . . .

[Pause, while image on screen shows injection pipette now aligned exactly in 3 o'clock position with respect to egg.]

but to do it with this sperm into . . . this . . .

[Does not finish the sentence as pipette penetrates the egg. MELANIE lets out audible gasp of relief.]

Felix

[Makes sudden start, as if he had been pricked.]

My God! You did it! Beautiful penetration!

[Image shows pipette resting within egg.]

Now shoot him out!

[Points to sperm head in pipette.]

Melanie

Here we go.

[Image shows sperm head at the very end of the injection pipette. She aspirates it back and gives it a second push.]

Ah, that's a good boy.

[Carefully withdraws pipette without apparent damage to the egg.]

Felix

You did it! Look at him, just look at him! Sitting in there.

[Approaches image and points to sperm head on screen. Calmer voice.]

It's amazing. That egg looks . . . what shall I say? . . . inviolate, almost virginal.

Melanie

[Looks up for first time from microscope.]

It better not be . . .

[Mimics his voice with a slightly sarcastic edge.]

'inviolate', to use your precious term . . . I violated it very consciously and tomorrow, I want to see cell division . . . Felix [points to VCR], *press the pause button, will you?*

[He does and image of fertilized egg remains frozen on screen in full view.]

Felix

So who is the father?

Melanie

There isn't any father in the usual sense of the word.

Felix

[Ironic]

An immaculate conception?

Melanie

You know, in a way that's true. There was no penetration of the woman, no sexual contact. In fact, at that moment, there was no woman, no vagina . . . nor a man [pause] . . . *The only prick* [pause] . . . *was the gentle one by a tiny needle entering an egg in a dish, delivering a single sperm.* [Laughs.] *Even that prick was provided by a woman. That process means nothing until the egg is implanted into the woman.*

Felix

And who is this woman? Whose eggs were you injecting?

Melanie

Mine!

Writing *An Immaculate Misconception* was hard work. The final version is the twenty-fourth draft! German, Swedish, French, Bulgarian and Japanese translations are already completed and some other languages are bound to follow. Contemporary plays are mostly watched, rather than read, yet I hope that the English, German and Swedish book versions of *An Immaculate Misconception*, already on the market, will be somewhat of an exception. They could well serve as a text book in which some key issues in contemporary reproductive biology could be played out by the students in some biology classes.

I had never written anything twenty-four times, but doing so with *An Immaculate Misconception* was an invaluable crash course in playwriting. I do not regret it for a moment, since it gave me the opportunity to meet and talk and work with so many different theater professionals. To some extent, the give and take of playwriting and play development resembles scientific research and publication, so different from the solitary focus of the fiction writer. These many drafts of my play were the result of a honing process that started in the fall of 1997 at a staged, rehearsed reading at the Tricycle Theater in London, followed by an opening at the 1998 Edinburgh Fringe Festival, and subsequent full theatrical productions in as diverse locations as London, San Francisco, Vienna, Vermont, Cologne, Sundsvall and Stockholm, culminating with a BBC broadcast in May 2000 as 'Play of the Week' on the World Service. When I listened to that radio broadcast, directed by Andy Jordan and performed by a stellar cast featuring Henry Goodman, Pennie Downey and Michael Cochrane, I felt reassured that my play had not been a misconception after all. The real bonus proved to be the German language premiere in Vienna. Not only was this the city where I had seen my first plays as a teenager, but it was held at the 100-year old Jugendstiltheater, which, coincidentally, was the site of the most erotic scene in my novel *Menachem's Seed*. No wonder that in my opinion, Isabella Gregor's production of *Unbefleckt* was the most imaginative of all the ones I had seen so far.

III

The relatively rapid acceptance of my first play can in large part be ascribed to the timeliness of the topic and the inherently dramatic aspects of human reproduction which in *An Immaculate Misconception* were presented so graphically—a feature commented on by all the reviewers. But as a chemist turning into a playwright, it behooves me to see whether chemistry can be presented as effectively on the stage as, say, sex. I had the good fortune to find a partner, Roald Hoffmann, interested in joining me in such a theatrical experiment (even though he is a theoretician by profession, rather than experimentalist). In 1981, while Professor of Chemistry at Cornell University, he was awarded the Nobel Prize in Chemistry for his theoretical chemical insights. But unlike most chemists, he has been interested for years in communicating with a broader public, and has done so through his own poetry and non-fiction writing.

Just as I tried in my first play to hide my didactic motivations behind the scrim of sex, in the second play, *Oxygen*, Hoffmann and I did this by taking up a theme—the Nobel Prize—that, at least to scientists, is potentially also sexy. The year 2001 is the centenary of the Nobel Prize; it is also the year in which our play is set. In *Oxygen*, the Nobel Committee decides to celebrate the centenary by establishing a new Nobel Prize, to be termed a 'Retro-Nobel,' to honor inventions or discoveries made before 1901, the year when the first Nobel Prizes were awarded. For a change, what's wrong with paying attention to the dead?

Our play attempts to deal with two fundamental questions: what is discovery in science and why is it so important for a scientist to be first? To put it even more crudely, why do the scientific Olympics only award gold medals and no silvers or bronzes? Why is it that in science, being second might as well be last? And yet, why in the end is it even more important to be recognized last? In *Oxygen*, we approach these questions as our imaginary Retro-Nobel Committee meets to select, first, the discovery that should be so honored, and then—as it turns out, not a straightforward question—which scientist to credit for it. Here is an early scene in which the chair of the committee, Astrid Rosenqvist, debates the issue with her male colleagues:

Astrid Rosenqvist

Now let me summarize whom we've got so far: John Dalton as father of the atomic theory . . . Dimitri Ivanovitch Mendeleyev for inventing the Periodic Table . . . August Kekulé for the structure of benzene, . . . and of course, Louis Pasteur . . . All of them first class . . . and a nice geographic spread: an Englishman, a Russian, a German, and a Frenchman—

Ulf Svanholm

And for a change no American!

Astrid Rosenqvist

Another advantage for concentrating on the nineteenth century. Or earlier. But we also agreed that they are candidates for a later Retro-Nobel. The first one must recognize where modern chemistry began.

Sune Kallstenius

In other words . . . with the discovery of oxygen.

Bengt Hjalmarsson

[Speaks slowly and pompously.]

'The Prizes should be distributed to those who have conferred the greatest benefit on Mankind.'

[Reverts to ordinary tone.]

That's what it says in Alfred Nobel's Will. Shouldn't that also apply to the Retro-Nobel?

Astrid Rosenqvist

Of course. But no one will question that oxygen confers great benefit on mankind, right?

Bengt Hjalmarsson

Oxygen was good for people before it was 'discovered.'

Ulf Svanholm

But there are plenty of benefits that require for oxygen to be isolated. What about the emphysema victim in an oxygen tent . . . the Everest climber with his oxygen bottles . . . the astronaut in the space suit?

Sune Kallstenius

We didn't pick oxygen for its value to mountain climbers or astronauts or sick people.

Ulf Svanholm

There you go with your usual spiel . . . the academic's ivory tower disdain for the useful . . .

Astrid Rosenqvist

Let's compromise. Who'd like to come up with some simple phrases to explain to the public that without the discovery of oxygen there would've been no chemical revolution... no chemistry as we now know it?

Bengt Hjalmarsson

I'll give it a try. Prior to Antoine Lavoisier—

Sune Kallstenius

You mean prior to Carl Wilhelm Scheele—

Ulf Svanholm

What about Joseph Priestley?

Bengt Hjalmarsson

Right back to the usual Nobel quandary! Too many candidates.

Throughout the play, as the Retro-Nobel Committee debates its selection, the audience learns about the three leading candidates through their wives in a trialogue that occurs in a Swedish sauna in 1777 just prior to a royal adjudication concerning the respective claims of their husbands: the Swedish apothecary Carl

Wilhelm Scheele, who made it first; the English minister-turned-chemist, Joseph Priestley, who published first; and the French chemist, tax collector, economist, and public servant, Antoine Laurent Lavoisier, who understood it first. In switching back and forth between 2001 and 1777, not unlike the time shifts in Stoppard's *Arcadia*, we present the historical and personal record that leads the Nobel Committee to its final conclusion. Without giving away the plot—after all, we want people to *see* our play—here is an excerpt from a 1777 scene:

Scheele

'Resolve the question: Who made fire air first?' That was His Majesty's command . . . to all three of us.

Lavoisier

Is that the real question?

Priestley

Of course. And you, Monsieur Lavoisier . . . did not make that air first . . . as you yourself in effect conceded yesterday.

Lavoisier

I understood it first . . .

Scheele

Understanding only comes after existence!

Priestley

But my dear Scheele! Proof of such existence must be shared!

Scheele

I shared it! Here is the letter . . . dispatched to you, Monsieur Lavoisier, almost three years ago. Describing work done earlier still.

Lavoisier

[Aggressive, yet words carefully chosen.]

I never heard about that letter until today . . .

Scheele

It describes the preparation of fire air . . .

Lavoisier

No such letter ever reached me.

Scheele

A recipe you reproduced in front of all of us today.

Lavoisier

But certainly not years ago as you now claim.

[Impatient]

What is the real purpose of this meeting?

Priestley

Priority! In August 1774 I made dephlogisticated air . . . your oxygen . . . at Bowood House . . . my patron's estate.

Lavoisier

Then you thought you had nitrous air, sir.

Priestley

The first steps of discovery are often tentative.

Lavoisier

Some of us are more careful than others.

Priestley

In October of that year, I met the leading chemists of France [Pause] *. . . including you, Sir.*

Lavoisier

Indeed, you dined in my house in Paris.

Priestley

I told the gathering . . .

Lavoisier

in your imperfect French . . .

Priestley

. . . which Madame Lavoisier comprehended fully . . . of my discovery.

Lavoisier

Your report . . . lacked clarity. Your methods were imprecise . . .

Priestley

Sir, your words are unworthy.

Lavoisier

At best, you, Dr Priestley, supplied us with the smallest of clues . . .

Priestley

I thought details mattered to you, sir.

Lavoisier

Only if they are relevant.

Priestley

More than once, my experiments in pneumatic chemistry were cited by you—

Lavoisier

Is that a reason to complain?

Priestley

Only to be then diluted . . . if not evaporated.

Lavoisier

How did I do so?

Priestley

You write

[heavy sarcasm.]

'We did this . . . and we found that.' Your royal 'we', sir, makes my contributions disappear . . . poof . . . into thin air! [Pause.] *When I publish, I say, 'I did . . . I found . . . I observed.' I do not hide behind a 'we.'*

Lavoisier

Enough of generalities . . . [Aside.] *. . . or platitudes.* [Louder.] *What now?*

Priestley

The question, sir! The question! Who made that air first?

Scheele

[Much more insistent than before.]

I did. I, Carl Wilhelm Scheele of Köping. And future generations will affirm it.

Priestley

But by the grace of God, I made it too . . . I, Joseph Priestley, and published first!

Lavoisier

They knew not what they'd done . . . where oxygen would lead us.

[The three men start arguing simultaneously in loud tones so that the words cannot be understood.]

[Tapping of staff.]

Court herald's voice

Three savants? Yet you cannot agree? So be it. [Pause.] *The king will not reward you!*

In contrast to the 24 drafts of my first play, Hoffmann and I needed only ten for our play to be accepted for radio broadcasts in 2001 by the BBC World Service as well as by West German Radio, and for theatrical premieres during the Nobel Centenary year by the San Diego Repertory Theater in California, a German premiere at the Stadtstheater in Würzburg and a

French one in Nantes. Even the book version of *Oxygen*—with blurbs by four Nobel laureates—has already appeared because the publisher shared our belief that this play can be 'read' as well as 'seen'. As an inveterate optimist—an indispensable trait for a playwright—I believe that both *Oxygen* and *An Immaculate Misconception* may gradually make it into the repertoire as repeatedly performed plays. And if they do, it will be the didactic features that will keep them there.

IV

Earlier on, I speculated as to why I had not encountered the names of any chemists among authors of plays dealing with science. Of all the reasons, the explanation that chemists are culturally imprinted not to use dialog in their writing seems to me the most persuasive. I thought that I might test that proposition experimentally with a very literary chemist, but the results, which I report here, are at best ambiguous.

Primo Levi is invariably and deservedly cited as a major literary figure of post-World War II Europe. Though not a researcher, he was an industrial chemist most of his life. He owed his wartime survival (as well as his most original and most admired literary work, *The Periodic Table)* to his chemical proficiency. I could not think of a better subject to test my hypothesis. My researches uncovered an ambiguous record. While dialog is essentially absent from the deeply personal accounts of his concentration camp experiences, that was to be expected, given their autobiographical focus. But since there is plenty of dialog in his four collections of short stories and his novel, *If not now, when?*, he clearly is not restrained by the monologist training I attribute to most chemists. So what about plays? How had he responded to the call of the stage? In the 1960s, when Levi's literary reputation first came to light, there is some evidence of thespian stirring. The Canadian Radio's broadcast adaptation in 1963 of *If this is a man* (a story of Levi's concentration camp experiences) stimulated him sufficiently that he himself prepared his own rendition in Italian and then followed it with a full stage version that was performed in Turin in 1966. (It has such a mob of actors that it is probably unplayable in these days of theatrical budgetary constraints.) But except for two science fiction playlets included in his 1966 collection *Natural Stories* (originally published under a disarming pseudonym, Damiano Malabaila) and a radio play of *The Truce* (an account of his return from Auschwitz) in 1978, the voice Levi's admirers mostly comment on is the reflective personal one of the chemist-writer.

To my regret, I never met Levi personally. If he were still alive, I probably would have first tried to ingratiate myself by commenting on his birth control story, *Small Red Lights*, before asking him whether he considered playwriting an avocation or just a minor diversion. But I surely would have regaled him with the tale of my greatest theatrical triumph, which, after all, happened right in his own country. In July of 1999, I was on my way to Spoleto to speak on 'Science on Stage' at the Spoletoscienza Festival, in response to an invitation prompted by *An Immaculate Misconception*. My wife was with me, because we planned to proceed from Spoleto to Sulmona, the birthplace of Ovid, on whose biography Diane was embarking. A driver was supposed to meet us at the Rome Airport, but once all waiting chauffeurs bearing placards with every kind of name but ours had departed, we were standing alone at the exit from the customs hall. Fifteen minutes must have passed, during which time my wife had gone to search for other transport while I manned the lonely gate in the hope that the driver from Spoleto might yet appear. At that moment, a woman pulling a heavy suitcase passed near me. When she saw me, she stopped. 'You are Carl Djerassi, aren't you?' she said, still panting. 'Yes,' I admitted, nonplussed, 'but how did you know that?' 'Oh,' she waved the question away, as if it were all too obvious. 'I saw your play at the Eureka Theater in San Francisco.' 'And you remembered that?' I asked flattered. That's when she punctured my balloon of pride by pointing to the next exit about 50 meters away. 'I did, when I noticed that man over there who's holding a sign with your name.'

'You see?' I would have told Primo Levi, 'I would probably have had to take a taxi all the way to Spoleto. Just think how much money I saved by writing that play.' I probably would not have stopped there, once I had started on the slippery slope of a playwright's braggadocio. 'I am now working on another "science-in-theater" play, called *Insufficiency*,' I would have continued. 'It's a catchy title, almost as good as *Immaculate Misconception*. Don't you think so?'

'It all depends,' I hear him say. 'What's it all about?' I'd shake my head. 'I can't tell you yet. So far, I've only written two scenes.' Primo Levi supposedly was a gentle, polite man, but I wouldn't have blamed him if he'd been irritated. 'So why are you asking me?'

Just guess.

Chapter 12

What if?

In our increasingly technocratic society, we seem to be much better at perfecting means than we are at deciding to what ends they should be used. Perhaps one reason for this difficulty is that we often confuse means and ends, thinking automatically that the possession of a technology determines its use. Techniques at either extreme of the reproductive spectrum—from preventing birth when it might occur to creating new life when it otherwise might not—are the most human examples of this dilemma. Much of the debate over contraception has assumed that means create ends—that access to contraception, for instance, invariably tempts otherwise chaste teens to copulate. If fear of pregnancy is the only justification for chastity, then in my opinion a practical, but not a moral or ethical case has been made for abstaining from intercourse. By the same token, advocates of human genomic research are sometimes surprised when people view the techniques arising from such research—such as screening for genetic disease or the insertion of a new gene—with suspicion, if not outright alarm. But do technocrats or technophobes alike credit the means with more influence on the ends than is actually justified?

In ancient Greece, whenever lawyers met after some interesting case, they teased each other with 'What if that had happened?' 'What if he had done . . .?' 'What if . . .?' Supposedly, that was the origin of fiction. But since my present account is not fiction—other than that all autobiographical musings contain elements of automythological fiction—let me end this book by asking what if the Pill had never been created? Posing that question is not pointless. I am convinced that if we had synthesized norethindrone on 15 October 1966 rather than 1951, and if no other organic chemist had appeared in that interval with a similar invention, oral contraceptives would not have existed in the year 2000. This is not the opinion of an arrogant chemist, however convinced he may be of the crucial role of his scientific discipline, but rather that of a realist. A 15-year lag in such chemical research does not seem all that implausible, considering that relatively few other chemical groups were then working on progestational steroids, in contrast to the enormous attention paid to anti-inflammatory corticosteroids. But such a delay in the research wouldn't have prevented the eventual confirmation that a steroid contraceptive pill was possible. Even if the chemical clock had started 15 years later, it would certainly have continued ticking, while Gregory Pincus or a similarly engaged biologist demonstrated in animals that such synthetic steroids could be used for what Ludwig Haberlandt had envisioned as oral 'hormonal temporary sterilization.' And there is little doubt that the watch would have ticked even louder as clinical researchers, John Rock or some later reincarnation, would have tried such steroids in humans. But, because hardly any pharmaceutical company with the financial, logistic, and marketing muscle necessary to bring oral contraceptives to the market would have been willing to carry on during the 1980s and 1990s, then the ticking most likely would have stopped. What are the reasons for this dismal assessment?

I

The Pill was born at the best possible time: in the early 1950s—the heyday of new drugs, but also a rather short window of opportunity. Pharmaceutical companies, the media, and the public proclaimed and accepted the benefits of the post-war chemotherapeutic revolution with barely a reservation. Every problem, be it a medical one such as tuberculosis or a social one such as the population explosion, seemed capable of a technological fix. By contrast, the 1960s proved to be the worst of times. The changed climate was triggered by the thalidomide tragedy, which resulted (primarily in Europe, where the drug was widely used) in the birth of hundreds of children with serious limb deformities to mothers who had taken that highly effective sedative in early pregnancy when the embryo is at greatest risk. That medical disaster raised everyone's consciousness of the importance of prior teratological (birth defect) studies in suitable animal models before introducing a drug that pregnant women might consume. While this would now seem self-evident, prior to 1960 it was not. To cite just one example: At high doses, aspirin is teratogenic in rats, mice, dogs, cats and monkeys, but hardly in humans, as demonstrated by a century of clinical use. But suppose it were introduced as a new drug in the early 1960s, when teratogenic tests had become obligatory. The birth defects produced after administration to pregnant animals would have raised such a red flag that few pharmaceutical companies would have been willing to pursue clinical studies in that thalidomide-sensitized climate with a drug that might be used widely—as aspirin indeed has been—by pregnant women.

A direct consequence of the thalidomide debacle was passage of the 1962 amendments of the Federal Food, Drug, and Cosmetic Act, commonly known as the Kefauver-Harris Amendments. These stipulated explicitly for the first time that in addition to safety, a drug's efficacy also had to be demonstrated—a requirement that had hitherto been only implicit in the FDA's mandate. In principle, all of these changes made sense, but the FDA was totally unprepared to oversee these new requirements, notably in the area of contraception. In the words of Peter B. Hutt, the former counsel of the FDA, 'For roughly the first ten years under that statute [the Kefauver Amendments of 1962], FDA basically was playing catch-up ball . . . a simple reflection of the fact that FDA did not have the resources when the statute was enacted, and was not given the resources following enactment . . . and there was no time whatever to think through the issues before decisions had to be made.' Hutt also admitted that 'FDA employees have been praised only for refusing to approve a new drug, not for making a courageous judgment to approve a new drug that has in fact helped patients and advanced the public health.' An example of these attitudes is the fact that for a ten-year period, the FDA refused to approve any new cardiovascular agent, because some agency personnel insisted that reducing hypertension would not necessarily reduce heart attacks, stroke, and kidney disease. Such drugs were introduced in Europe years before they finally reached American consumers; some critics of the FDA claimed that the delay had caused at least fifty thousand unnecessary deaths and crippled over two hundred thousand people annually. In such a regulatory climate, contraception is especially vulnerable, because it is generally considered to be the practice of a healthy person, whom regulatory agencies (and society, for that matter) are not prepared to expose to the kind of risks that might be tolerated in an individual suffering from a disease. For contraceptive research especially, extreme caution became the FDA's watchword in the 1960s. The result was a predictable, if enormous, increase in the cost, and even more important the time, required to reach FDA approval. But worst of all, the pharmaceutical industry was faced with the loss of a significant portion of the 17-year patent life of a drug and hence its proprietary position by undertaking such prolonged studies. Under these conditions, only drugs capable of commanding extremely high prices over their shortened patent life would be likely to return what the industry (though perhaps not society at large) would consider an acceptable profit.

An even stronger disincentive became the escalation in America of legal liability suits. If chemical work on orally effective progestational steroids had only been completed in 1966, under the more stringent FDA requirements, advanced clinical studies would probably have been initiated only in the mid-1970s. This was precisely the time when the biggest contraceptive disaster, easily matching the thalidomide tragedy, had caught the public's and the media's attention. In 1970, the A.H. Robins Company, a fairly conservative, mid-sized American pharmaceutical company, had acquired the rights to a new IUD design developed by Dr Hugh J. Davis, then of Johns Hopkins University. Called the Dalkon shield, it was touted as superior to all other intrauterine devices. Since medical devices, unlike drugs, were barely registering on the FDA's radar screen at that time, the Dalkon shield was put on the market within months, and was soon chosen by hundreds of thousands of women for birth control. By 1974, it had become clear that Dalkon shield users had started to suffer from increased rates of serious infection, permanent infertility, and in the worst cases death. A veritable legal explosion ensued, with over 6000 liability suits eventually reaching multi-billion dollar levels; these were eventually elevated to class-action status, and ultimately caused the bankruptcy of the company in 1985. Ironically, the Dalkon shield's developer, Hugh Davis, had been one of the most virulent opponents of the Pill during the 1970 'Nelson Hearings' of the US Senate devoted to the side-effects of the Pill, claiming that his IUD was a far safer alternative. Which only goes to show that assessment of risk is itself a risky business.

The final nail in the coffin of this imaginary belated Pill would have come in the late 1970s, as the international pharmaceutical industry came to center increasingly on blockbuster drugs. This focus should not have been surprising, of course, given the pressure the new FDA regulations had put on profits. But the exact nature of the mix of drugs that came out under this new emphasis—the psychopharmaceuticals, anti-inflammatory, cardiovascular drugs, and anti-cancer chemotherapeutics—revealed the influence of another set of social, economic, and political trends: the emphasis on diseases of an increasingly geriatric patient population, the only age-group in the country to have achieved universal health insurance. A 1988 survey of the international pharmaceutical industry showed that new human fertility control methods were not even among the top 35 on their list of research priorities—well below such socially and medically essential items as antihistamine nose drops! I consider this market judgment logical but also tragic, because history demonstrates, in capitalist as well as in socialist countries, that no major advance in drug innovation can occur without the active participation of the pharmaceutical industry—in production, distribution, development, and even research.

It should be clear by now that the pessimistic conclusion I draw from my hypothetically delayed female Pill applies in spades to the prospects of a Pill for men. A limited number of reproductive biologists and clinicians are currently interested in male contraception using steroids that inhibit sperm production, coupled with the administration of testosterone to maintain libido, but none of that interest is likely to induce a major pharmaceutical company to undertake the gamble of bringing such a Pill into a drugstore. (In the year 2000, none of the ten largest pharmaceutical companies in the world is active in male contraception.)

So let us pretend that at the start of the twenty-first century steroid oral contraception for women is still a scientific research curiosity. That scenario clearly would please some critics, ranging from feminists such as Germaine Greer on the left to religious and social fundamentalists on the right. To Father Richard Welch, the author of the statement that 'The smallest, most innocuous seeming step—a tiny little pill—can cause endless damage,' I could then say, 'Don't worry. That tiny little pill does not exist.' Yet he would have found plenty of 'damage' at the start of such a Pill-less new century.

Timothy Leary and other recreational drug gurus; the Beatles and other rock and roll musicians; the hippies and the flower children—the social and cultural icons of the 1960s—would still all have left their indelible marks. And with them, make no mistake, we would have had greatly increased sexual freedom, leading, in spite of possibly increased condom, diaphragm and IUD use, to a plethora of unwanted pregnancies. Some of them would have been hidden from the Pill's social critics through the time-honored palliative of the shotgun wedding, but I have no doubt that the number of back-alley abortions would have escalated to such an extent that the legalization of abortion would actually have arrived in America a couple of years earlier than the *Roe* v *Wade* Supreme Court decision. Simone de Beauvoir, Betty Friedan, Germaine Greer, Gloria Steinem and the other powerful flag bearers of the woman's liberation movement had already been born. They all would have written the words that launched the most important social revolution of the past few decades. It was the women's movement coupled with more effective birth control (and not the other way around as posited, for instance, by the anthropologist Lionel Tiger) that unlocked the doors of the kitchens, nurseries and bedrooms so that women streamed into the outside working world with all the consequences that so many Pill critics bemoan and, of course, blame on the Pill. In 1900, 6 per cent of American married women worked outside the home, compared to the present

61 per cent (or 64 per cent of married mothers with children below the age of 6). Why ignore one of the major triggers of the shift that antedated the Pill by two decades: World War II, when millions of women entered the labor force and tasted the fruit of economic independence?

The HIV/AIDS pandemic would probably have arrived on precisely the same week, because infections caused by illegal drug use and promiscuous homosexual intercourse were not started in the laboratory of the synthetic steroid chemist, but rather in our personal laboratory, the brain and endocrine glands that have been around for millions of years. Of course, we must not be so taken up by the political and moral qualms of the Pill's critics that we forget the Pill's most important and (dare I say?) intended, effect, its influence on the reduction in the growth of the world's population. As I have already stated earlier, the dramatic drop in the birth rates of Japan and Europe cannot be credited or blamed on the Pill. I would credit the Pill in part with the more recent drop in birth rates of those lesser developed countries where the Pill has become the preferred method of birth control, but I suspect that even many of the critics would applaud the end to which the technical means have been put in this instance.

Just consider the fact that among married women in Latin America, oral contraceptives are the second most widely practiced method of contraception after sterilization. Brazil (the world's fifth largest country) alone has six million Pill users. Asia is another telling example. China relies primarily on IUDs and sterilization for birth control, but it still has the largest number of married Pill users (7.6 million) in the world. Or

Bangladesh, the eighth largest country in the world, now has 21 per cent Pill users (compared with 3.3 per cent in 1983), thus making oral contraceptives overwhelmingly the most popular form of birth control. North Africa contains some instructive examples: 31 per cent of married Moroccan women are on the Pill (compared with only 14 per cent in 1980) in contrast to the 1.4 per cent and 4 per cent incidence of condoms and sterilization, respectively. Algeria, with 44 per cent married women on the Pill, currently holds the developing world's record. There is no doubt that in those countries, with relatively high maternal mortality rates, tens of thousands of deaths associated with childbearing were prevented by the Pill. (A point that often gets left out of discussions of the Pill's safety is the risks of pregnancy itself—even in developed countries these are not trivial, as for instance in life-threatening ectopic pregnancies, and in most cases these hazards are greater than the risks of using the Pill.)

Until recently, sex and reproduction were inexorably linked: couples could only conceive through copulation. But that coupling of sex and fertilization is now heading for divorce and the question may well be asked whether the Pill is one of the causative factors. In one sense it is, because the Pill has made it very convenient to indulge in coitus without worrying about reproductive consequences. But the Pill is innocent when it comes to the technological implementation of fertilization in the sex-fertilization relationship.

II

The technical developments in assisted reproduction, starting with artificial insemination and ending with various *in vitro* methods, had nothing to do with the Pill. Deciphering the human genome and all the other dramatic developments in medical genetics, including the spectacular advances in genetic screening methodology, would have arrived at exactly the same time—Pill or no Pill. The use (or misuse) to which such genetic screening will be put is a colossal problem, with huge moral, ethical and social implications. But that again would be with us without the Pill, and indeed is with us already through the increasing use of amniocentesis or chorionic villus sampling to determine if there is anything 'wrong' with the developing fetus. What we are facing now are the ramifications of moving from *fetal* screening techniques to pre-implantation *embryonic* genetic analysis—a technique that more than any other will cause *fertile* people to turn to the IVF techniques that have so far been developed for the treatment of infertility. Fetal screening is to many people an after-the-fact approach with the only alternative then being abortion. Pre-implantation genetic screening, however, offers a choice among several embryos and most couples choosing that path do not equate discarding an early embryo created in a petri dish to an abortion. *I believe that it is that embryonic screening possibility associated with IVF that will lead increasing numbers of fertile couples to resort to that technology.* If there ever is an example of the means complicating the ends, this is it in the field of human reproduction, because such an approach bothers many people whose opinions must be taken into consideration in any pluralistic society.

Of course all of us are willing in one way or another to glamorize or demonize new reproductive technical advances: the Frankenstein myth has been with us for too long. I can think of no better example than the artificial placenta. Work in that field started in the 1950s, around the time of (but unrelated to) the early Pill development, when various investigators, using sheep fetuses, attempted to see whether extrauterine fetal incubation was feasible. That research, abandoned in 1980, has been taken up again during the past ten years in a much more sophisticated way by Nobuya Unno in Tokyo. In his experiments, four-month-old goat fetuses were removed from the uterus, catheterized to maintain proper blood circulation via a centrifugal pump and maintained in an artificial glass placenta for up to three weeks. While no live goats have been born so far, the progress achieved so far has been remarkable. Unno is a practicing clinician, specializing in neonatal obstetrics, whose clearly stated goal it is to develop life-support systems for human neonates unable to sustain their lives without such heroic intervention rather than implementing some Frankenstein scenario. Preliminary evidence suggests that such artificial placentas may eventually enable clinicians to have 22-week-old newborns survive.

Whether this is medically or even ethically appropriate is a separate and very serious question. But when I witnessed a public lecture, with video illustrations of actively moving goat fetuses in a glass container, the response of the public and especially the media was unfortunately revealing. They focused entirely on the possibility of complete extra-uterine reproduction: fertilizing an egg under the microscope and then transferring the embryo into an artificial environment and eventually an artificial placenta—the ultimate 'test tube baby.' Questioners ignored the monstrous technical problems to be overcome:

moving from an embryo through all the intermediate states to a fetus presents unimaginably complex technical barriers (how, for instance, do you catheterize a creature that changes its form from blobby tadpole to human, and its size by several orders of magnitude, all in a matter of weeks?), not to mention the socio-economic questions such a ruinously expensive pregnancy would raise. None of which seemed to faze the audience to this talk: they only wanted to know when all this would be accomplished so that a woman, who wishes to have her own genetic child without going though a pregnancy could do so by placing her fertilized egg rather than prematurely born child into such an artificial womb. The fact that the few women who do not *wish* to go through such a pregnancy (or the many more who cannot do so because of a hysterectomy) could proceed today with very little risk by making arrangements with a surrogate mother at a fraction of the multi-million dollar cost of such an artificial pregnancy seemed to be of little interest.

Which brings me back to the Pill. The dismal conclusions I draw from my 'what if?' alternative history would imply that the Pill played a negligible role during four decades of continued clinical use. If all these things would have happened anyway, why did we bother with the Pill in the first place? Why did millions of women in the developed world accept the Pill in the 1960s with totally unanticipated speed as their preferred choice for contraception? Why, at the end of the twentieth century, is the Pill the top modern contraceptive method among married women in 78 out of 150 countries and the most widely used form of contraception worldwide (if the figures for China and India are omitted)? Of course, the Pill played an enormous role, partly causative, though mostly facilitative, by the simple fact that at precisely the right time in Western social history, a convenient oral contraceptive became available that completely divorced contraception from sex. The fact that a woman could do so in privacy and that it was her decision alone was crucial. But what were the costs associated with such a conscious choice? For many religious persons, notably Catholics, the moral choice was serious, because it led millions of women to a conscious and continuous act of disobedience against explicit religious teaching. While some people applaud the resulting secularization of the Church, clearly many deplored it and still do so. The biological cost took much longer to dissect.

III

No drug category has been studied with such intensity and in terms of so many parameters as have steroid oral contraceptives—from the effects on ear wax production to a possible carcinogenic role. Tens of thousands of articles and numerous books have been published on such clinical scrutiny, and any summary by me would only be a collection of second-hand information. Furthermore, my book is a summary of utterly personal reflections and not a 'how to' birth control manual with all its associated caveats. But in any 'what if?' inquiry, it is necessary to touch upon the subject of side-effects; without such a discussion, such a scenario is necessarily (perhaps dishonestly) incomplete. I have already commented on the wishy-washy definition of 'safety' in a medical context. In contrast to the romanticized dictionary definition of absence of side-effects—how often do we hear the wish for a 'perfectly' safe contraceptive?—

let us accept immediately that the Pill, like any drug, has side-effects. And as always, these need to be divided into acute and long-term ones. For instance weight gain or nausea may appear to be negligible to some, but many a woman may consider them serious enough to make the Pill unacceptable. But the major established health risks are certain circulatory system diseases. Thromboembolic disorders, for instance—the formation of blood clots in the circulation, with potentially disastrous results—clearly fall into that category. In the 1960s, these were found to be associated with the amount of the estrogenic rather than the progestational component of the Pill, and were exacerbated in heavy smokers and older women. Eventual reductions of the dosages of estrogen led to a greatly diminished occurrence of such events. A somewhat elevated incidence of strokes (thrombotic and hemorrhagic), though rare, is another established circulatory system risk factor. But even here, consider the difficulties in establishing quickly such dangers when the absolute number of cases is low: in late 2000, after 50 years as a common ingredient in a variety of non-prescription cold and cough remedies and appetite suppressants, phenylpropanolamine has suddenly been associated with an increased risk of hemorrhagic stroke in younger women.

By far the most serious and complicated side-effects are long-term ones, of which the question of potential tumor production, of course, was and still is foremost in people's minds, as it should be whenever a drug is to be consumed over long periods. Even now, four decades after the Pill's introduction, epidemiologists debate whether its prolonged use increases the risk of breast cancer. Many women will be discouraged to learn that a totally unambiguous answer can still not be provided, because the dosages of both the progestational and the estrogenic components of the Pill have been progressively lowered since the mid-1970s, whereas tumors take years and decades to grow. The long-term studies are only telling us about the potential effects of *yesterday's* Pills, which only confirms an insufficiently appreciated truism about long-term side-effects: *the absence of evidence for side-effects is not evidence for the absence of side-effects*! But pooled epidemiological evidence in 1996 from 54 studies in 25 countries covering over 150,000 women (one-third with breast cancer) suggests that women who had used oral contraceptives within the past 10 years were somewhat more likely to be *diagnosed* with breast cancer than non-users. It is still not clear whether the Pill promotes the development of an already existing tumor or whether women on the Pill have more frequent breast examinations, thus leading to a finding of more tumors at an earlier stage. This argument is supported by the observation that duration of Pill use has no effect on risk and that such diagnosed women have fewer cancers that spread beyond the breasts. In summary, for those who are suspicious of the Pill solely because it is synthetic and 'not natural,' let me remind them that Nature is not benign. So why assume that 'natural' is synonymous with 'safe' or that 'synthetic' is inherently dangerous or bad?

Any discussion of side-effects needs to bear in mind an often-neglected truth: not all side-effects are bad. Some readers may have heard of Zyban, a drug marketed to help smokers break their addiction to nicotine; this life-saving effect was discovered as a side-effect to the drug's original formulation, when it was known as Welbutrin and marketed as an anti-depressant. Or the widely prescribed drug Zocor, which now seems to strengthen the bones of those often-elderly patients who consume it to lower their cholesterol. An interesting 'what if?' question applies to the beneficial, *non-contraceptive* effects of the Pill. Some have been known for a long time, especially menstrual benefits: reduced dysmenorrhea and premenstrual symptoms, more regular menstrual cycles, and lowered iron deficiency anemia (especially important in undernourished poor women) due to lighter menstrual bleeding. Another category of noncontraceptive benefits was uncovered during the long-term epidemiological studies looking for negative side-effects. I am particularly referring to protection against ovarian and endometrial cancers —an unexpected positive feature that was first noted among users of relatively high dosage Pills in the mid-1970s. If there had been no Pill, thousands of women would have died of these diseases during the intervening decades. More recent studies have demonstrated that such protection is also provided by the newer lower-dosage steroid regimens. Protection against benign breast disease is more dosage dependent, with higher amounts of the progestational component of the Pill offering greater protection. (Fibrocystic breast disease was half as frequent in women taking the 2.5 mg norethindrone acetate Pills compared with consumers of the 1.0 mg variant). And finally, evidence is accumulating that bone mineral density is positively impacted by oral contraceptives in women approaching menopause or suffering from eating disorders, who otherwise are more likely to display symptoms of osteoporosis. In these cases, the protective function is provided by the estrogenic ingredient of the Pill.

But there is one categorical answer that can be given to the question of the Pill's benefits versus its risks, and that one is so overpowering that it alone justifies the continued existence of the Pill. I am referring to the numbers of unwanted children and abortions—the sheer mass of human suffering—prevented over the years. The critics who discount these are precisely the ones who ascribe in an oversimplistic way the deterioration of the nuclear family to sexual philandering prompted by the Pill. Lionel Tiger in *The Decline of Males* went even further, ascribing the progressive diminution in male entitlements to the Pill, because 'for the first time in human experience, and perhaps in nature itself, one sex is able to control making babies.' (Plenty of women are likely to ask Tiger why, since only one sex can make babies, that sex should not also make the ultimate decision.) One of Tiger's arguments for the reduction of male reproductive power is that with the Pill, sneaky women can now seduce men into becoming unwilling fathers by claiming that they were on the Pill while purposely deciding not to take it. All of these critics neglect the reverse proposition: that great sex often leads to stable marriages; that the quality of sexual congress in a loving couple is enormously increased when the fear of an unwanted pregnancy is removed; and that children who are wanted are more likely to be loved.

I conclude with the most personal of 'what if?' questions: what would have happened to me if the Pill had not existed? As I showed in possibly too-excruciating detail, I would have been a very different human being. Still the same organic chemist, still a seeker for 26 hours in a 24-hour day, but otherwise a very different man: not a fiction writer nor a playwright; probably not an art collector; most certainly not the teacher I have turned into during the past three decades—in summary a much less socially engaged individual. If I were a neutral observer of that Carl Djerassi, especially were that observer a woman, I probably would have turned away. No wonder I am grateful, deeply grateful, to this man's Pill.

Index

abortion 81–7, 103, 104, 120
abortion pill 84–6, 191, 192
abstracts 7, 212-13
acetylene substituent 47
Acht Uhr Abendblatt 18
Adams, John 151, 237
aging 174, 286
Akademik Oparin 160
Algeria 289
'already now' 139–42
Alston, Alfred N. 212
Amirkhanian, Charles 239
Arcadia (Stoppard) 250–3
Arlecchino (restaurant) 223
aromatization 44–6
Ars Poetica (Horace) 245
art collection 216–28
Art in an Age of Mechanical Reproduction (Benjamin) 259
artificial insemination 118, 121, 132, 133
artist's colony 236–42
Asbell, Bernard 15
Auspices (29 July 1977) (Middlebrook) 141–2
authorship of papers 209–12
Available Light (Adams) 237

Bangladesh 289
Baulieu, Etienne-Emile 84, 191
Beauvoir, Simone de 69, 287
Benadryl 54
Benjamin, Walter 259
Berggruen, Heinz 220–1
Berkeley Barb 95–7
Biography of the Drug that Changed the World, A (Asbell) 15
'Biosocial Aspects of Birth Control' course, Stanford 194–201
biotech 179

Birch, Arthur J. 46
birth rate 68, 82–3
Blake, William 3
Blinded by the Sun (Poliakoff) 245–48
Bloch, Felix 234
bone mineral density 296
Boone, Charles 237
Born, Gustav 17
Boston Women's Health Book Collective 94
Botanicamex 27
Bourbaki Gambit, The (Djerassi) 170–6
Brave New World Revisited (Huxley) 63–4
Brazil 288
Breaking the Code (Whitemore) 253–4
breast cancer 294–6
Brown, Louise Joy 121–2
Butenandt, Adolf 35

cancer
- anti-cancer drug, putative 104
- breast 294–6
- cervical 51–2
- of Djerassi 152
- endometrial 296
- ovarian 296

Cantor's Dilemma (Djerassi) 156–8, 161–2, 163–4, 168–9, 205, 209–10
case histories 249
Castor's Dilemma (Djerassi) 150, 155–6
Catholicism 116, 117, 293
Center for Research on Women (CROW) 235
cervical cancer 51–2
Chang, Min-Chueh 52, 59
chemists and chemistry 188
Cheuse, Allan 237
Childs, Lucinda 237
China 100, 130, 288
cholesterol 30, 33
Christies' auction 223
CIBA 36
Clampitt, Amy 240
Clock Runs Backward, The 7–10
cloning 134
'Cohen's Dilemma' (Djerassi) 149–50
coitus interruptus 98, 99
Coleridge, Samuel Taylor 249
Colton, Frank 52, 53, 55, 56, 57
Coming upon an unidentified work of outdoor art near Woodside, California (Clampitt) 240
condoms 118
- dispensers at Stanford 203–4
- Djerassi's collection 204

contraceptive
- choice, national differences 98–101, 288–89
- research 74–5, 76–7, 89, 192

Control of Fertility, The (Pincus) 20, 56
Copenhagen (Frayn) 246, 254–7
corpus luteum 17, 18
cortisone 35–6, 37, 39–41, 42
cystic fibrosis 124

Daily Mail 79, 95
Dalkon Shield 99–100, 285
Davis, Hugh J. 285
Davy, Humphrey 248, 249
Decline of Males, The (Tiger) 297
Dempsey, E.W. 19
Dennis Rosen Memorial Lecture 248
developing world 93–4, 289
Die hormonale sterilisierung des weiblichen organismus (Haberlandt) 17
Dioscorea 21, 22
diosgenin 21, 23, 33–4
Diosynth 27
Djerassi, Alexander Maxwell 7
Djerassi, Carl
  academic career 36–7, 151, 189, 193–204, 205–13
  accent 139
  accident on ranch 145–6
  art collection 216–28
  artist's colony 236–42
  cancer 152
  at CIBA 36
  condom collection 204
  fiction writing 145–50, 153–8, 163–7, 170–87, 221
  financial gains from Pill 95–7
  Himalayan trek 151–2
  Inhoffen Medal 48
  Marker interview 22–8
  National Medal of Science 192
  personal life 137, 151
  PhD thesis 44
  playwriting 124–9, 244, 246, 258–78
  poetry 137–8, 142–4
  reading tours 157, 169
  Santa Cruz land ownership 228
  Sunday Times Magazine Top Thirty persons list 4
  at Syntex 37–9, 46, 95, 189–90, 192, 216
  wedding 151
  on White House 'Enemy list' 234
  Wolfson interview 11–12
Djerassi, Dale 158–61, 228
Djerassi, Norma 229
Djerassi, Pamela 228, 230–4, 243
Djerassi Foundation 233–4, 236–8
Djerassi Resident Artists Program 238–42
dogs, sperm preservation 131
Dramamine 54
drugs, package inserts 74

Edwards, Robert 122
Ehrenstein, Maximilian 35, 43–4, 48
Ehrlich, Paul 194
Elsen, Albert 235
embryonic development 181
embryonic genetic screening 134–5, 290
emergency contraception 89–2
Endocrine Laboratories Inc. 13
endometrial cancer 296
Enovid 53
Epel, David 181
Ephron, Nora 145
estradiol 44, 45
estrogens 44–5
estrone 44
'Ethical Discourse through Science-in-fiction' course, Stanford 205–13
Ethyl Corporation 28
Eugenic Protection Law 103
extrauterine incubation 291

FDA 283–5
Federal Food, Drug, and Cosmetic Act 283
Feininger, Lyonel 221
Feminine Mystique, The (Friedan) 69
feminism 69–71, 78–80
'Feminist Perspectives on Birth Control' course, Stanford 202–4
Fenwick, Jean-Noël 246
Fernholz, E. 35
fiction writing 145–50, 153–8, 163–7, 170–87, 221
Fieser, Louis D. 39
First-class Nun (Djerassi) 182
Ford Foundation 193
Fraenkel, Ludwig 17
Frankfurter Allgemeine Zeitung 161–2
Frayn, Michael 246, 254–7
'French death pill' 85
Freud, Sigmund 18
Friedan, Betty 69, 287
Fukuyama, Francis 113
Futurist and Other Stories, The (Djerassi) 155, 221

Garcia, Celso-Ramon 59–60, 62
Gehry, Frank O. 237
Geist, Del 237
German 66–7
Gideon Richter 18, 27, 86
Greenblatt, Robert 51
Greep, Roy O. 61
Greer, Germaine 79, 286, 287
Gregory Pincus Memorial Lecture 60
Grosz, George 221

Haberlandt, Ludwig 16–19, 20, 56, 59, 281

Hamburg, David 194
Hapgood (Stoppard) 249–50
Harper's Magazine 40
Hawking, Stephen 255
Headington Hill Hall 153
Heartburn (Ephron) 145
Hechter, Oscar 61
hemophilia 130
Hertz, Roy 16, 51
Hewitt, Cecil 65
Himalayan trek 151–2
Hoechst 85
Hoen, E. Weber 212–13
Hoffmann, Roald 268, 276
Holland 98
'honorary' authors 209, 210
Horace 245
hormonal temporary sterilization 16
hormone replacement therapy 44
Hormosynth 27
Huang, Minlon 49
Human Biology Program, Stanford 193–204
Human Life International 87
Human Sum, The (Rolph) 65–6
Hutt, Peter B. 283
Huxley, Aldous 63–4

Immaculate Misconception, An (Djerassi) 124–9, 259–68
immortality 114–36
impotence 177, 179–80
In the Carolinas (Stevens) 140
in-vitro fertilization (IVF) 92, 122, 123, 290
India 130
Infecundin 19
infertility 51, 122–4, 177
Inhoffen, Hans Herloff 35, 45, 47, 48, 49–50
Inhoffen, Peter 49
Inhoffen Medal 48
Innsbruck University 16
Institute of Medicine, 76
intracytoplasmic sperm injection (ICSI) 123–30, 260–2
isomer 54
Italy 68, 98
IUDs 99–100, 109, 285

James Irvine Foundation 237
Japan 68, 101–13, 288
Jitsukawa, Mariko 102
John D. and Catherine T. MacArthur Foundation 237
Johnson & Johnson 55
Judaism 116, 117–18

Kandinsky, Wassily 221
Kefauver-Harris Amendments 283

Kennedy, David 14
Kennedy, Donald 194
Kharasch, Morris 28
Klee, Felix 221
Klee, Paul, collection by Djerassi 215–16, 217–23, 226, 227
Kosei-sho 103–4, 105, 108–9
Kretchmer, Norman 194

Laboratorios Hormona 23
Landscape near Bear Gulch Road (Lewis) 239–40
Latin America 289
Laurentian Hormone Conference 60–2
Lederberg, Joshua 194
Lehmann, Federico 23–4, 33
Leibo Stanley 132
Leighton, Patricia 237
Leonardo da Vinci 2
Levene, Phoebus 28
Levi, Primo 277–9
Lewis, Janet 239
liability suits 75–6, 284–5
Life 40
Life of Galileo (Brecht) 250–1
Lingonberries Are Not Sufficient (Djerassi) 138
Lipschutz, Alexander 51
litigation 88
Lodge, David 244–5
Loma Prieta earthquake 157
McCormick, Katherine 15
McCue, Gerald 229
Magdalen College, Oxford 153
Mainichi Shimbun 108, 112
Makepeace, A.W. 19
Maldives 158, 161
male contraceptives 76, 78, 177, 286
Marker, Russell 20–30, 33, 34–5, 41
Martyn, Rhonda 151
Maruyama, Hiromi 102
Marx, deceased (Djerassi) 182–7, 224
Maryland University 28
Maskenfreiheit (Djerassi) 182
Maxwell, Betty 153
Maxwell, Robert 7, 153
Mead, Margaret 14
'Medicine 256' course Stanford 205–13
Menachem's Seed (Djerassi) 177, 178, 259, 261
menstrual disorders 51, 52, 101–2
Merck and Co. 39
Mexico
  Dioscorea collection 21–2
  pharmaceutical industry 13, 23–5, 27, 33, 42, 58
Middlebrook, Diane Wood 138–42, 146–7, 151, 153, 155, 163, 231, 232–3, 235, 236
'Middles' 145, 146–7, 182

mifepristone (RU-486) 84–6, 191, 192
Miramontes, Luis 47, 50, 95
Mislow, Kurt 170
moral decline 80
'morning-after Pill' 90–2
Morocco 289
Morris, John M. 90
Mosher, Harry 28
Mössner, Ursula-Maria 168–9
Murray, Herbert C. 42

Nash, David 240–2
National Academy of Sciences 164–5
National Medal of Science 192
'natural family planning' 117
Nature 132, 209, 212
Nelson hearings 73–4
New York Times 111–12, 156
Newsweek 40
Newton, Isaac 3
Nilevar 55
nitric oxide 179, 180, 181
NO (Djerassi) 177, 178–82, 210–11
norethindrone 13, 47–8, 50–2, 53, 54–5, 96
norethynodrel 52–4, 56
Norlutin 52
norprogesterone 43, 46
nortestosterone 55
North Africa 289
Novel as Communication, The (Lodge) 244

Ogino calendar rhythm method 103
Old Goat (Hoen) 212–13
O'Leary, Dennis 239
O'Shee, Kriss 237
osteoporosis 296
Ota ring 109
ovarian cancer 296
Oral contraceptives
  health benefits 76,
  bias in male researchers 13–14
  'morning-after' 90–2
  non-English terms for 66–8
  package inserts 74
  side-effects 71–4, 293–6
  societal effects 69–95
  use statistics 4, 288–9

Pincus, Gregory 15, 16, 20, 51, 52, 53, 54, 56, 59, 60, 92, 281
Place, Francis 99
placenta, artificial 291
playwriting by Djerassi 124–9, 245, 247, 258–77
poetry by Djerassi 137–8, 142–4
Poliakoff, Steven 245, 247–248

Politics of Contraception, The (Djerassi) 163
polymerase chain reaction (PCR) 176
Population Council 15
population growth 11, 93–4
Portable Stanford series 163
postcoital Pill 90–2
post-menopausal pregnancy 133–4
pre-implantation screening 134–5, 290
Princeton University 170
progesterone 19, 21, 23, 24–5, 26, 27, 33–4, 35, 41–2, 43, 44, 51–2
Pro-Life 81
public policy 190–2, 195, 204–5
Pugwash Conferences 178
Pure Scientist, You Look With Nice Aplomb (Djerassi) 143

Raymond, A.L. 57
reading tours 157, 169
religion 86–7, 116–19
Renga 208–13
Rice-Wray, Edith 60
Rickey, George 236
Robins, A.H. Company 285
Rocco's (restaurant) 224
Roche 58
Rock, John 20, 54, 59, 62, 281
Rockefeller, Sharon 201
Rockefeller Foundation 15, 16, 199–200, 201
Rockefeller Institute 28
Rolph, C.H. 64–6
Romania 82–3
Rose, Paul L. 257
Rosengart, Angela 219
Rosengart, Siegfried 219
Rosenkranz, George 34, 37, 47, 50, 95
Rothen, Alexander 29
Roussel-Uclaf 84, 85
Royal Institution 248–9
RU-486 (mifepristone): 84–6, 191, 192
Russell Marker Lectures, 29

safety 71
Sagan, Carl 164
San Francisco Chronicle 214
San Francisco Museum of Modern Art 214–16, 226
Sanger, Margaret 14–15, 69–70
sapogenins 21
Sarrett, Lewis 39
Schering A.G. 45, 47
Schläfriger Arlecchino (Klee) 223
Science 162, 190, 191
science-in-fiction 149, 165–7, 170, 205–6
science-in-theater 245–79
'Science Renga' 208–13

Searle, G.D. & Company 52, 53, 54, 55–6, 57, 59, 101
Second Sex, The (Beauvoir) 69
Sei-ai: Sei no nai kankei 106
sex predetermination 130
Sexton, Anne 155
sexual intercourse 116–21
  separation from reproduction 4, 121
Shikibu, Murasaki 1–2
Shionogi & Company 101
Shipley, Elva, G. 13–14, 15–16, 48
Short, Roger V. 119–20
Sic manebimus in pace 234
side-effects 71–4, 293–6
Simon, Leon 56–7
Simpson, O.J. 176
Slotta, Karl 19
SMIP 228, 234, 236, 239
Somlo, Emeric 23, 24–7, 33, 57
Sotheby's auction 221–2
sperm 116, 118
preservation 130–1, 133
Stanford University 151, 189, 193–204, 205–13
Steinem, Gloria 287
Steirteghem, André van 123, 260
stereoisomers 44
sterilization 92, 98–9, 132, 133
steroids
  biological activity 32
  defined 30–2
  total synthesis 32–3
Steptoe, Patrick 122
sterols 30–1
Stevens, Wallace 140–1, 142, 151
Stoppard, Tom 249, 250–1, 252–3
stroke 294
strophanthidin 43
structure elucidation 29
sukimono 107
Sunday Times Magazine 1
Sylvan Steps (Nash) 240
Syntex 13, 23, 25, 27, 33, 34, 37–9, 42, 46, 54, 56, 57–8, 95, 189–90, 192, 216
synthesis, partial/total 30

teenage pregnancy 98
testosterone 34, 45, 47
Teutsch, Georges 84
thalidomide 282–3
Three variations on a theme by Callosobruchus (Djerassi) 144
thromboembolism 294
Tiger, Lionel 81–2, 287, 297
Time has come, The (Rock) 20
To the Roaring Wind (Stevens) 141
Top Thirty persons of the millennium 1–3, 4
total synthesis 30
'Toyota Cantos, The' (Djerassi) 154–5
Tyler, Edward 51, 101

Ulmann, André 84
Understanding Modern Poems (Middlebrook) 139
United Kingdom 98
Unno, Nobuya 291
Upjohn Company 41–2

vaccines 88
Van Wagenen, Gertrude 90
vas deferens 123
vasectomy 132, 133
'Vatican roulette' 117
Vaughan, Josefa 237
Viagra 108–9
Vienna 158–60
Vietnam War 234
Vladivostok 160–1
'Vocalissima' 141

Wang, Shirley 204
Washington Post 225
Weber, Bruce 249
Welbutrin 296
Welch, Father Richard 86–7, 95, 286
What's Tatiana Troyanos Doing in Spartacus's Tent (Djerassi) 147–9
White House 'Enemy list' 234
Whitemore, Hugh 253–4
William and Flora Hewlett Foundation 237
Wilson, Harold 153
Wisconsin University 44
Wolfson, Jill 11–12
womb, artificial 292
women in the workforce 287–8
Worcester Foundation for Experimental Biology 15, 52, 61
Worlds into Words: Understanding Modern Poems (Middlebrook) 163

Yamamoto, Izuru 144
Year of the Fire Horse 110–11
Young, Neil 228, 230
Yu, Jennifer 204
Yuzpe, Albert 90, 91

Zaffaroni, Alejandro 54, 101
Zoecon Corporation 189–90
Zyban 295–6